Process Engineering with Economic Objective

Advisory Editor: Paul Butler, PhD
Editor, *Process Engineering*

A guide to

Process Engineering with Economic Objective

G. L. WELLS
University of Sheffield

LEONARD HILL   An Intertext Publisher

Published by
Leonard Hill Books
a division of
International Textbook Company Limited
24 Market Square, Aylesbury, Bucks. HP20 1TL

First published 1973

ISBN 0 249 44116 0

Printed in Great Britain by
Billing & Sons Limited, Guildford and London

To Joan

Contents

PART 3   THE OVERALL PROCESS, OPTIMIZATION, AND
         SIMULATION

List of Figures

Nomenclature

Symbol	Name
a	case identifier
a	input state
a	region start
A	cash value
A	hydrogen flow
A	product identifier
A	surface area
b	case identifier
b	input state
b	region end
B	better case
B	cash value
B	fired heaters
B	fuel gas flow
B	product identifier
B.F.W.	boiler feed water
b.h.p.	brake horse power
B.T.X.	benzene, toluene, and xylene mixture
c	cold stream
c	input state
C	compressors
C	conversion/pass
C	cash value
C	operating costs
C	product identifier
C	specific heat
C	capacity
C.I.	cast iron

Symbol	Name
C.W.	cooling water
d	depreciation, fraction of investment
d	input state
D	converters and towers
D	cash value
D	diameter
D	distillate flow rate
D	distillation stage
D.A.	optimization algorithm
D.C.F.	discounted cash flow
E	exchangers
E	design parameters
E.F.D.	engineering flow diagram
f	function
F	tanks
F	cost factors
F	feed rate
F	function
F.C.	flow controller
F.G.	fuel gas
g	development grant
G	pumps
G.P.	gradient projection
h	hot stream
h	difference between numbers
h	premium to cover risk
h.p.	horse power
i	rate of interest on investment
I	capital cost
I	investment
I.C.	incremental change
k	enhancement factor
k	years
K	keep
l	length from region boundary

Symbol	Name
L	length
L	reflux flow rate
L.	linear
L.C.	level controller
L.P.	linear programming
L.P.G.	low-pressure gasoline
LR	length of plate-bending rolls
m	maintenance costs
M	optimum state
M	throughput/annum
M	total state variables
M.A.P.	optimization algorithm
M.A.R.	minimum acceptable return
N	number of stages
N	sets of independent design equations
NC	number of circumferential plates
NP	number of logitudinal plates
N.P.V.	net present value
N.L.	non-linear
o	overhead costs
p	element, usually pivotal
p	probability
p	royalties, fraction of investment
p	state of stream
P	parameter vector
P	penalty
P	pressure
P	profit per annum
P.C.	pattern change
P.F.D.	process flow diagram
P.P.N.L.	optimization algorithm
P.R.C.	pressure-recording controller
q	decision variable
q	thermal state of feed to distillation unit
Q	capacity
r	royalties fraction

Symbol	Name
R	reaction stage
R	radius
R	ratio
R	realization
R	reflux ratio
R	reliability, successes/number of trials
R	replace
R	gross return
R	production rate
S	flow rate
S	money in fund
S	sensitivity coefficient
S	separation stage
S	total sales realization per annum
Sa	salvage value
S.O.V.	shut-off valves
t	taxation
t	temperature
t	time
T	temperature
T	trials
U.	economic objective function
U	overall heat transfer coefficient
U	total costs before taxes
V	overhead vapour rate
V	venture profit
V	volume
w	weighting
W	flow rate (distillation bottoms)
W	venture worth
W	worse case
W.H.B.	waste heat boiler
x	mole fraction
x	variable
y	variable
Y	objective function

Subscripts	Name
z	probability factor

Subscripts	Name
A	case identifier
b	base
b	benzene
B	base case
c	cold
c	column
c	cooling fluid
c	manufacturing costs
ci	coolant in
co	coolant out
conv	converged
d	direct cost factors
D	distillate
D	overhead
e	ends
E	delivered
e.c.	economic criteria
f	input
f	probability of failure
F	installed
F	cost factors
F	feed
F.G.	fuel gas
h	hot fluid
H	hydrogen
HK	heavy key
i	indirect cost factors
i	interest rate
i	component and stage number
i	installed cost
ii	component and stage number
if	stage feed

Subscripts	Name
io	stage output
I	reactor identifier
I	capital cost
I	investment
k	enhancement factor
k	years
K	constant
K	cycle
K	years
ln	logarithm
L	liquid
LK	light key
m	minimum
m	minimum acceptable return
M	maximum
M	throughput/annum
n	project life
n	number of stages
n	parameters
N	time
o	overhead
opt	optimum
p	parallel
p	project
r	row
R	reactor
R	recycle
R	reflux ratio
s	sales income
s	series
s	sides
s	speed
s	vapour velocity
t	time
T	total

Superscripts	*Name*
U	objective function
v	vapour
v	variable
V	venture profit
w	working capital
W	bottoms

Superscripts

$-$	statistical mean
m	factor
$\circ$	pure
$'$	differential
$'$	distinguish between cases
$*$	optimum

Greek

α	relative volatility
δ	partial differentiation operator
η	efficiency
θ	time
θ	Underwood's constant
λ	latent heat of vaporization
ϱ	density
σ	variance
ϕ	objective function

Script

$\mathscr{T}$	transform

Preface

This guide to process engineering with economic objective is intended as a guide rather than a review; to serve as an introduction and to integrate the fuller instruction available from specialist texts. Emphasis is on the overall process and its economic objective and for a projected chemical plant this is the simulation of the economic balance sheet. This entails integration of various factors which have been classified in this book as process economics, process specification, and process optimization and simulation; or as shown in the structural diagram.

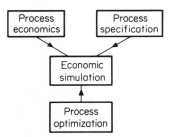

Much of the contents are used in a second-year course in chemical engineering at Sheffield where the aim is to introduce the student to chemical engineering practice. Initially emphasis is placed on economic objective so that the student appreciates that engineering calculations must lead to economic reality. Also the extent of such calculations may be limited by the objective function. At the same time I would not like any student to feel that he is motivated only by whether his work is economic. This would curtail his technical interests and lead to a duller working life.

If the economic objective is fully appreciated it can radically alter effort on any project. In particular it affects the engineering content at the capital evaluation stage. It has always seemed illogical to place a lot of effort into achieving accurate capital cost estimates when value of production is inexact.

In the chapters on process economics, one aim is to show how the economic criteria used must frequently be tailored to the specific design problem and only the person carrying out the design is able to select the appropriate financial procedure. Also the time value of money does affect one's conclusions.

Process charts represent a section of process engineering activities seldom included in the literature. Yet they are so vital to the practising engineer. On any project virtually all the chemical engineers on the project, from design to commissioning, will study the engineering flow diagrams and usually they will be able to suggest improvement. The charts show the importance of displaying information clearly and in the required detail and lead to a study of the techniques for evaluating mass/heat balances and specifying equipment.

The use of specification sheets in student design problems is strongly recommended. They need not be fully completed. They add a touch of professionalism to the project and make the results of sizing a column more than a completed tutorial sheet. Also process calculations should result in specification of items such that costs of appropriate accuracy may be obtained. Furthermore the optimum size can often only be obtained by cost comparison of different schemes. There is a tendency to evaluate a parameter as being the optimum factor and this may prove false. For instance a reactor may be designed as the minimum size of bed but to achieve the appropriate operating conditions requires higher costs than for rejected schemes. Similarly such sub-optimization may result in the overall process equipment costs increasing. Care should be taken to study the overall system and its optimization.

At present considerable effort is going into producing mathematical models of plants both for design and operation. It is desirable that these always incorporate a costing mode. Obviously once capital is expended the model must adapt and more accurate cost information on aspects such as maintenance, reliability, etc., be incorporated. However, a desirable aim is to produce a model complete with process economics at the design stage, followed by modification at the production stage and by subsequent integration into the corporate model of the company.

The book is essentially a guide and it is hoped that process technologists of different backgrounds will find it useful. Programs and computational techniques have been omitted from this text, but will be considered in a companion volume, *Computation for Process Engineers*, to be published by Intertext later this year.

G. L. Wells
Sheffield 1973

Part 1

PROCESS ECONOMICS

Chapters 1-5

The first part relates the objective function of the design to the extent of the process engineering required. It is commonly assumed in the literature that all process engineering calculations eventually result in a chemical plant, whereas in practice the majority of schemes are never installed. As financial information is introduced it is important to know the accuracy to be achieved, i.e. there is no point in achieving great engineering accuracy in a preliminary estimate.

Economic criteria and plant balance sheets are discussed. It is emphasized that financial equations should be integrated within process calculations to achieve the optimum design. Some typical examples are given of small systems treated in this way. No attempt is made to present a comprehensive list of optimization techniques. The main aim is to show that a costing mode should preferably be added before an optimum is evolved.

In some examples throughout the book British rather than S.I. units are used. This has been done to permit ease of comparison with the earlier work referenced in a given example.

1 The Objective Function of the Process Engineer

1.1 The role of the process engineer

The role of the process engineer varies according to his employment. His objectives are the same as those of his company and are primarily economic.

1.2 Chemical company objectives

A chemical company exists to sell chemicals at a profit in a competitive market. It is essentially a manufacturing organization. To continue in existence the company must continue improving its performance in existing marketing areas, develop new products for these markets, take advantage of market growth and penetrate new marketing fields.

To meet these objectives the company is organized into the operational sections of production and sales supported by technological, research and development, engineering, and purchasing all serviced/controlled by personnel and finance. The task of the production department is to produce chemicals in quality and quantity for delivery as required by sales. All resources should be used as efficiently as possible, and are identified on the production balance sheet at costs appropriate to decreed standards. At the same time materials must be produced safely and with due regard to the environment. Every attempt is made to improve performance and to seize opportunities for new processes, etc. The engineering department maintains such facilities and provides a project engineering service. The technological department provides a process engineering service. This includes the process design of plant, preliminary assessment of new projects in conjunction with research and development, and a technical service on existing plants. An active role during the commissioning of new plant is usually incorporated in the service. Finally, research and development is concerned mainly with new processes or assistance in some problems in the manufacturing area.

1.3 Chemical contractors' objectives

Although some chemical contractors also manufacture chemicals or equipment and vice versa, basically the contractor provides a project engineering

3

service to a chemical company with the aim of making a profit. Such service is provided as appropriate to the type of contract under which the contractor is employed. The organization of the chemical company is geared to act through the project leader on individual projects and is well indicated by Rase and Barrow (1957). Some contractors specialize as licensors of a particular process which they guarantee for throughput and yield; generally this applies for only a short period but obviously their reputation would be low should their plants be unsuccessful.

1.4 The basic objective function

The basic objective function of the process engineer within these different roles is economic. Although solving technical problems he must never lose sight of this objective. Thus the basic strategy of the process engineer should permit the adjustment of the technical aspects of the design to produce the most profitable solution. At the same time any company should appreciate that to achieve this objective function due consideration should be paid to motivation. Unfortunately the work of the process engineer is invariably judged in economic terms and cannot fall into any alternative category, i.e. an architect might claim that a non-functional building is an art form. However, he should be encouraged in carrying out some exercises which are of genuine interest to him, even if the company pay-off is negligible. If this develops the individual's knowledge and creative ability all to the good. After all a later project may benefit by the improved performance of the engineer. Furthermore he should be encouraged in that, although naturally he is disappointed if some work is discarded, as an individual he has developed from accomplishing this task.

2 The Extent of the Design

2.1 Stages for capital authorization

As the objective function of any process engineering exercise is almost always economic, then it is logical to relate the accuracy to which the economic information is required and the extent of the process design. Usually such economic information is related to a proposal for expenditure of money, so the objective function initially aims at giving an investor an indication of the possible profit from a given capital outlay. The investor may then decide whether the opportunity is worth while. The process engineer is attempting to give the investor sufficient information for a decision to be made, yet making an effort to minimize design costs on abortive projects. Once a favourable decision is made, the process engineer maintains the profitability of the project above an acceptable level by balancing capital and operating costs and by ensuring the process can achieve the planned throughput. A further objective is introduced in that the capital expenditure should be accurately assessed; the economic environment results in most investors requiring an accurate estimate of the amount of capital they need to spend.

Furthermore, for a contractor this is his main objective function. Obviously he would never obtain work if the investor did not decide to spend money, but once this is assumed his profit is frequently associated with capital expenditure. Of course profitability of the process after installation influences the future business which the contractor might obtain.

The extent of the design depends not only on the size and complexity of the unit but also on the stage which it has reached for capital authorization of the project. These stages are clearly indicated by a study of the usual practice in estimation of capital cost (*Capital Cost Estimation*, 1969). Each scheme, once approved, is followed by the next stage. The stages are as follows.

Preliminary. A 'should the project go ahead' stage before any formal application for capital expenditure is made; accuracy $\pm 30\%$ for capital cost including a contingency.

Budget. This may comprise several stages in which the accuracy of the estimates are refined. However, essentially it is a stage in which the go ahead for the project is sought; accuracy $\pm 10\%$ for capital cost.

5

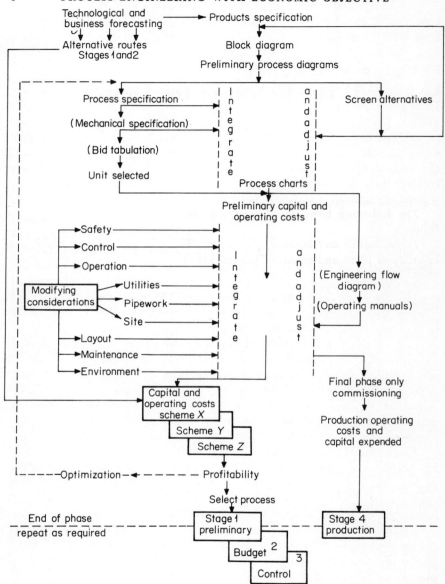

Figure 2.1 Systematic chart for process engineering (items in parentheses do not start until stage 3, control)

Control. An accurate capital cost, $\pm 5\%$, is required for control of expenditure during the construction period.

Production. Any plant passing through the control stage will go into production. As additional personnel become involved in process engineering aspects it is convenient to introduce this as a stage of detailed checks. The final capital cost represents the actual expenditure.

2.2 General description of the stages of capital expenditure

Let us examine how the objectives fit these stages and describe them further. First though, we introduce some terminology which is amplified later in the guide.

A systematic chart. To assist in the breakdown of the stages of capital authorization a planning chart (Figure 2.1) has been developed. It follows the main design activities. Strictly it is necessary to have four charts, one for each stage, but as each stage is slightly different for each project I consider it better to indicate the overall pattern and let experience dictate the detail.

The following terms have been used in the chart and are defined as follows.

Technological and business forecasting. These procedures determine the capacity of plant and extent of the market. They provide input information about the plant to the process engineer.

Process charts. Process charts describe the process. The block diagram indicates the main plants in a complex, or the main items in a plant. The process flow diagram shows the items of equipment, the majority of process streams and control systems, and contains information on the process materials and physical state. Engineering flow diagrams give further detail on streams, valves, and control.

Process specification of equipment, The principles of chemical engineering are used to size the various units and to indicate all nozzle connections.

Mechanical specification of equipment. Mechanical specification of equipment involves upgrading the process specification with information such as thickness of metal and codes of practice to be used, so that the vessel can be manufactured.

Modifying considerations. Certain aspects such as control, maintenance, site, environment, etc., although they affect aspects of the design of the process units, are essentially modifying constraints on the capital and operating costs of the plant.

Profitability and optimization. As the objective function is economic the process engineering exercise ends only at this point. Optimization exercises cycle through the profitability stage and back to the design of the equipment. They are carried out as and when required and not at the end of the stage. The profitability exercise can take several forms and depends on the extent of information available. It aims at indicating to an investor the return on the capital which he is being asked to expend.

2.2.1 PRELIMINARY, STAGE 1

This is not solely a development stage although it could be the logical output from a research and development department. It represents the process engineering required to indicate whether a further exercise to achieve capital authorization should follow. It should incorporate business and technical forecasting. In some cases the proposed plant need not be itemized. However, unless the plant is a standard package the preliminary process charting exercises should not be omitted, particularly for a complex of plants. Exercises should be incorporated for the screening of alternatives. These include method study, systems engineering, or even 'brain-storming' techniques and 'think' tanks. This stage gives opportunity to the creative ability of process engineers. Not only is this beneficial for the project and the company but it should be stimulating for the personnel concerned.

As any project at this point is uncertain to be installed, the amount of detail, and hence design costs, should be limited. Often capital and operating costs for an existing plant can be obtained from contractors or publications. In this case the main 'modifying considerations' applied are company costs. If process design is necessary, use short-cut procedures and existing computer programs wherever possible. The cost of equipment should be obtained from factorial procedures or company records, the values for individual units combined and overall factors used to obtain the capital cost of the plant.

2.2.2 BUDGET, STAGE 2

This stage normally determines whether the project will go ahead. First consider the objective function. Theoretically this is profitability, but any company is concerned about capital expenditure and thereby introduce a second objective. As in general larger inaccuracies arise in profitability estimates from wrong prediction of throughput, value of production, and yield, than from energy requirements and capital costs, making an objective a capital cost accurate to $\pm 10\%$ means more design work is required than solely for the profitability exercise. One can accept this for a contractor. If he does not produce an accurate forecast, then depending on the contract he will lose either goodwill or money, and his company profit is associated with both. But a chemical company should in theory be prepared to make capital available for sound projects. Thus, if the project profitability estimate were sound, why worry (within reason) about capital? Unfortunately chemical companies are concerned, and, although money would be funded if overexpenditure occurs, the need for such an application often reflects badly on the staff concerned, even though their profitability forecast was perfect. This results in an unfortunate overemphasis on capital costing.

For complete plants alternative tenders may be sought and in this respect the budget estimate may closely resemble a control estimate. For equipment competitive tenders are not sought but quotations of costs may be obtained. This results in errors as on a bid-tabulation considerable deviations can ensue. Thus there is small point in designing internals at this stage and mechanical specification should be strictly curtailed. Similar considerations arise with regard to schedules for instrumentation, electrical, buildings, and structures. A general rule is only to consider in detail anything so special as to warrant additional effort. Thus a similar project to one tackled previously may be estimated by updating earlier information. A special project may have considerable information available from preliminary studies.

To achieve the required accuracy means a full assessment of the site. Contractors may get over this problem by stating that their budget quotation relates to a specific load-bearing characteristic of soil. Before submitting the estimate the company has to revise the cost to the required accuracy.

All these factors add to the design costs. If a capital cost accurate to $\pm 15\%$ were accepted for budget estimates factorial methods could be used, thereby reducing costs on discarded projects. The principles of cost estimation should be tailored to the needs of the various stages of design development.

2.2.3 CONTROL, STAGE 3

The plant is now going to be installed. The process charts are analysed and refined to give capital cost savings and optimize energy requirement. Detailed process and mechanical design is carried out and specification sheets sent to manufacturers.

In many cases information may be incomplete and left for the manufacturer to provide. Alternatively, previous designs can be used to provide this information. Tenders are received and bid-tabulations evolved. Project engineering commences in earnest and an accurate capital cost to within $\pm 5\%$ is produced. This is the main economic objective at this stage. Modifying considerations play an important part, decisions are made regarding extent of instrumentation, structures, etc., and a reliability analysis is carried out to ensure that production can achieve throughput. It is essential to ensure the process will deliver the required quantities of products in accordance with specification under all foreseeable conditions. Decisions to change or omit units, etc., should be based on profitability. This the chemical company must ensure. Obviously it depends on the type of contract, but the contractor is more concerned with achieving the budgeted capital costs and reliability of the design. He must also keep his design costs down. The objective of the chemical company is to maximize profitability of the plant. The contractor's is to achieve a satisfactory profit on the job. The two aspects are not completely interrelated. Even in the chemical company, within which theory suggests emphasis on profitability is predominant, some personnel adhere rigidly to the minimum capital cost as the criteria to use in selecting items. This is understandable. After all companies exert great pressure to keep the capital within limits and yet do not check so rigidly any increase in operating costs. A further understanding of the objective function is required at corporate level. This will be brought about by uncertainty analysis of financial data and emphasis on design with economic objective at this stage of the project.

2.2.4 PRODUCTION, STAGE 4

As indicated earlier, a separate stage is termed production. Although it is at the control stage that initial engineering flow diagrams are produced, they are subject to much checking and even after obtaining a control estimate they should continue to be reviewed. Safety checks (Fawcetts and Wood, 1965) can then be fully implemented by the operating companies' personnel. It is useful to go through a full systematic design exercise, this stage being important for carrying out detailed checks. These include a study of product specifications and manufacturers' drawing. Considerable savings in commissioning time can be achieved by process engineering work on the plant items as they are erected. Thus minor errors in, say, liquid hold up on a tray might be corrected, as a visual check on the actual item sometimes conveys information better than on the drawings. Also it is useful training for those process engineers involved in the commissioning. Carrying out design calculations familiarizes them with the plant theoretically as well as visually. The profit statement at this stage is of course the actual plant performance and is presented as a plant balance sheet upon which initial plant standards are based. The capital cost represents actual expenditure which is checked against the previous estimate.

In theory some part of the contingency of the project should be left to cover minor modifications but in practice this is exceptional. Note that profitability as here expressed has no meaning once expenditure has been carried out.

3 Accuracy of Financial Data

Financial information of varying accuracy comes into the project. The accuracy of capital cost estimation has been mentioned. How accurate is other information and what effect have inaccuracies on the profitability? If the company require a budget estimate accurate to $\pm 10\%$, the process engineer must provide information which fits in with this accuracy. But he must base his decisions on profitability. It is emphasized that accuracy of capital cost and profitability are not the same.

3.1 Accuracy of capital cost estimates

The accuracy of capital cost estimation in relation to project development is defined below. In addition an estimate of capital cost is required for optimization exercises.

Preliminary estimate. This estimate is accurate to about $\pm 30\%$ with a large contingency (15–20%).

Budget estimate. This estimate is required to assess the viability of the proposed process and long-term financial planning. It is required to an accuracy of $\pm 10\%$.

Tender estimate. A more accurate version required to $\pm 15\%$ replaces a budget estimate in a 'turn-key' quotation and project.

Control estimate. To control expenditure in the construction stage an accuracy of $\pm 5\%$ is required.

For an estimate to be reliable requires a clear definition of its extent and program, together with a sound contract containing provision for escalation and efficient project management. In addition a contingency must be included.

3.1.1 PRELIMINARY ESTIMATE

Few problems exist in achieving the preliminary estimate. Values for the typical distribution of costs (*Capital Cost Estimation*, 1969) can be used to assess the overall capital cost once a rough idea is obtained for the equipment

11

cost. For example, given the location of a proposed system, a rough sketch of the process flow sheet and approximate sizes of the major items of processing equipment, a competent engineer can in favourable cases assess the investment to an error of less than 15% (Hackney, 1965). Delivered equipment may be divided into towers; drums and tanks; pumps; compressors; heat exchangers; special equipment; heaters and stacks. Include insurance, taxes, and overseas shipping to 'free on board' prices for imported equipment. Equipment costs may be estimated using factors in correlations such as

$$I = I_B \left(\frac{Q}{Q_B}\right)^m$$

where I is the investment, Q is the capacity, subscript B denotes the base case, and m is a factor, typically (0·5 to 0·7). A cost index is required to relate costs for different years to the same basis. Note that there are economic limitations on the capacity to which equipment can be scaled. A range of values are published regularly in the technical press. To obtain the installed cost from the delivered cost of a unit, factors are added for piping; structures; electrical; instruments; fire-proofing; building; foundations; sewers; site development; insulation; painting; contractor's overhead; profit and costs; royalties; engineering fee and a contingency, exactly as for the overall costs. As an indication of order of magnitude Hand (1958) recommended overall factors as follows.

Fractionating columns, pressure vessels, pumps	4
Heat exchangers	3·5
Compressors, miscellaneous equipment	2·5

The above breakdown into centres of activity for a project is used together with costs of equipment for control of costs in subsequent stages.

Numerous breakdowns of costs occur in the literature. For any unit or indeed plant it is possible by suitable use of factors to obtain an expression for the installed cost, Table 3.1.

3.1.2 BUDGET ESTIMATE

The problem at this stage is to achieve the ±10% accurately with limited expenditure on design costs and time. The equipment cost for a budget estimate can be estimated using factors or graphs showing the price of equipment varying with capacity. The tonnage of metal is often used to assess costs. Alternatively, as in Chapter 8, one can build up the cost from a schedule of rates.

Often budget quotations are obtained by contacting a single manufacturer or company records. However, no matter how detailed the design, errors are introduced by not going out to competitive tender. This may result in considerable deviations and I have a bid-tabulation showing a multi-stage

Table 3.1. Analysis of capital costs for chemical plant construction (within limits)

	Liquid processing								Solid/liquid	
Case	1	2	3	4	5	6	7	8	10	11
Total cost £000's	200	300	100	50	200	235	2400	Major	80	250
Remarks	Plant in open with instrument control room	Existing building modified	New building	Existing building	New process building including warehouse	Plant in open with small buildings	New plant site	Estimate from Perry	Existing building	New warehouse building
Equipment	23	35	29	32	21·5	27	39·9	23	52	31
Installation of equipment	1	1·5	1	2	1·5	2	—	—	6·5	2
Buildings	8	12	32	—	36	3·5	1·6	8	—	19
Foundations, etc., excluding buildings	3	2·5	8	3·5	1	2·5		3	2·5	1·5
Structures	7	2·5	3	4	2·5	5	3·9	8	6·5	1·5
Pipework	14	14	12	23	10	14·5	14·3	13	6·5	9
Electrical	7	4	4	4·5	4	7·5	8	8	3·5	6·5
Insulation and paint	3	5	1·5	1·5	3	3·5	3·5	5	1·5	2
Instruments	12	4·5	1	5	2·5	17	4·8	6	0·5	2·5
Engineering	22	19	8·5	26	15·5	19	21·5	26	20·5	25
Totals (%)	100	100	100	100	100	100	100	100	100	100
Offsites							25	24		

pump varying in price from £2600 to £6000. Even for equipment involving no complex internals prices vary. This in part relates to variation in overhead and design charges. Capital costs are greatly affected by the contractor's feel for the job, the amount of work in the office, and how much he wants the job and the type of contract. The latter is considered later but the other aspects have a great effect on the effort which goes into the estimate, and the extent of overheads and profit to be included in the estimate. This commercial side of the estimate can only be learned by experience.

As outlined in Chapter 2 similar considerations apply to other schedules. For further information see texts on capital cost estimation (*Capital Cost Estimation*, 1969; Bauman, 1964). I.C.I. claim that their FACTEST program, when costs of main plant items are known accurately, produces estimates of total plant cost within $\pm 10\%$. As the name implies this is by a factorial procedure based on their records. Every job presents different considerations and it is virtually impossible to give guidance apart from building up the cost and modifying the extent of work using commercial judgement.

3.1.3 CONTROL ESTIMATE

The refinement of the design and the availability of bid-tabulations results in the ready estimation of the capital cost. Process and items should be optimized using economic design criteria. Of particular importance is to avoid selecting items of equipment solely on the basis of minimum capital cost. This statement should be noted by operating companies who suffer the increased operating costs that ensue. Bid-tabulations should be analysed fully, see Example 4.7.

Apart from balancing operating and capital charges, compare installed costs. There is a tendency to buy standard units and an exchanger may be split on the bid-tabulation into two smaller units at a lower price. However, this is the equipment cost. The installed cost complete with additional piping, etc., may not show to the same advantage.

To ensure the estimate proves accurate, limit costs to those authorized and focus cost control efforts where most effective. The project manager must have overall responsibility for the estimate and must be able to maintain a close scrutiny of cost returns. He must plan his commitments to prevent the contingency sum being expended too rapidly.

3.1.4 DESIGN COSTS AND CONTINGENCY

Design costs will vary with the contract and thus the form of bid. Some typical examples are illustrated in Table 3.2. The list is not comprehensive. Licensing fees may be required in addition to the contractor's overheads and profit. If the design is carried out by the chemical companies' own staff, then similar charges should be made against the project. This assists in showing the economic viability of those departments involved.

A contingency is required to cover estimating errors, omissions and unforeseen factors, design data changes, and escalation. It varies according to the contract; a higher contingency would be felt necessary for a lump sum contract. A risk analysis is carried out on aspects of the capital cost. Thus if a new catalyst were being used the contingency might be increased to reduce possible loss due to failure. If the project contains new process development a larger contingency, say, 15% should be used than if there is a degree of repetition of previous experience when, say, 5% should be used. Escalation on site charges which the contractor might foresee should be allowed for within the contingency. However, if such escalation is due to taxation, this should be covered by clauses in the contract. To guard against escalation due to delays in approval of expenditure, all estimates should be firmly linked to a time schedule.

Table 3.2. Some typical methods of bidding

Type of bid	Overhead profit	Design	Purchasing services	Materials	Construction
open cost	%	open	%	open	open
fixed fee and open materials	fixed	fixed	fixed or %	open	open
materials fixed after design is completed	fixed	fixed	fixed or %	fixed	open
unit rates	fixed or %	unit rates	unit rates	open	unit rates
lump sum	fixed	fixed	fixed	fixed	fixed

3.1.5 WORKING CAPITAL

An allowance must be made for working capital which is the capital in the form of ready cash required to meet operating expenses, inventories of stock, etc. One month's operations is a typical figure.

3.2 Uncertainty in financial data

The above considerations with regard to the accuracy of capital cost estimates are based on convention. However, accuracy of estimates should be related to the accuracy of the financial data. The design objective function U may be taken as a function of the parameters E_1, E_2, E_3, ..., E_N.

$$U = U(E_1; E_2; E_3; ...; E_N)$$

If this objective function is computed using values of $\bar{E}_1$, $\bar{E}_2$, $\bar{E}_3$, ..., $\bar{E}_N$, representing the statistical mean from a range of data, this gives a value of the objective function $\bar{U}$; one can measure the dependence of U on small changes in the variable parameters. The sensitivity coefficient is defined as the fractional change in the objective function, divided by the fractional change in the parameter which caused such change in the objective function. For a small change ΔE_1 in parameter E_1 the sensitivity coefficient S_1 becomes

$$S_1 = \frac{\text{(objective function after change in } E_1) - \text{(objective function before change in } E_1)}{\text{(change in } E_1)}$$

or

$$S_1 = \frac{U(\bar{E}_1 + \Delta\bar{E}_1, \bar{E}_2, \bar{E}_3, ..., \bar{E}_N) - \bar{U}}{\Delta E_1}$$

In the limit this becomes

$$S_1 = \frac{\partial U}{\partial E_1}$$

and similarly for $S_2, ..., S_N$.

The likelihood of the parameter E taking on a value in the range E to $E + \Delta E$ will follow some form of probability distribution function $p(E)$. It is simplest to assume the data are encoded in a normal distribution, which in normalized form is

$$p(E) = \frac{1}{(2\pi)^{1/2}\sigma} \exp\left\{\frac{-(E-\bar{E})^2}{2\sigma^2}\right\}$$

where $\bar{E}$ is the mean or most probable value of E and σ^2 is the variance or mean square deviation

$$\sigma^2 = \frac{\sum_{i=1}^{n}(E_i-\bar{E})^2}{n}$$

The relationship between the uncertain parameter and objective function is usually non-linear. But, by approximation to linearity, the analysis of the propagation of uncertainty is simplified. Rudd and Watson (1968) show that, if the uncertain parameter E_i follows a normal distribution with mean $\bar{E}_i$ and variance σ_i^2, then so does the uncertainty in the objective function U, with a mean

$$\bar{U} = U(\bar{E}_1, \bar{E}_2, \bar{E}_3, ..., E_n)$$

and variance

$$\sigma_u^2 = \sum_{i}^{n} S_i^2 \sigma_E^2$$

Note that often a full range of data is not available and such information may be simply the managers' optimistic, pessimistic, and best guess estimates. A normal distribution curve reflects the managers' opinion that, for instance, sales will not fall below his lower sales limit or go above his upper estimate. As the relative likelihood is greatest near the centre and diminishes near either end-point, this reflects his general feeling of the projected figures.

In a normal distribution the probability of E falling in a given range is the integral of the distribution over that range. Important values are the probability of E falling within $\bar{E}\pm\sigma$ which is 0·683, within $\bar{E}\pm2\sigma$ which is 0·954, and within $\bar{E}\pm3\sigma$ which is 0.997 (Figure 3.1). A study of such a distri-

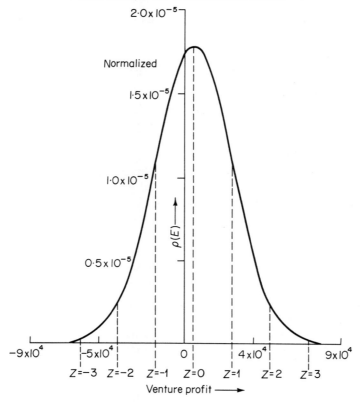

Figure 3.1 Distribution of venture profits

bution might suggest that it is 95% certain that the real value of the parameter is within the pessimistic and optimistic estimates. Then doubling to allow for ±

$$4\sigma = \text{optimistic} - \text{pessimistic estimates}$$

Statistical tables are available which give the value of the integral based on the standard form of the distribution equation

$$p(E) = \frac{1}{(2\pi)^{1/2}} \, e^{-z^2/2}$$

where $z = (E - \bar{E})/\sigma$ and the curve is in normalized standard form (whole area under the curve is unity). The tables refer to one-half of the distribution only, so the probability of E falling in the range $\pm z$ is given by twice the value of the table. Thus, knowing the mean $\bar{E}$ and the deviation σ, it is possible to plot the distribution for various estimates of E.

Example 3.1. Uncertainty in sales and capital estimates. To show the effect of variation in sales and capital estimates on the profitability analyse two systems.

Case 1. Constant selling price.

$$V = (50{\cdot}8M - 12M - 0{\cdot}15I)(1 - t) - 0{\cdot}1I$$

Case 2. Variable selling price.

$$V = \underset{\text{sales revenue}}{(76M - 0{\cdot}0036M^2} - \underset{\substack{\text{variable} \\ \text{costs}}}{12M} - \underset{\substack{\text{fixed} \\ \text{costs}}}{0{\cdot}15I)} \underset{\text{taxation}}{(1 - t)} - \underset{\substack{\text{minimum} \\ \text{acceptable} \\ \text{return}}}{0{\cdot}1I}$$

where M is the items/year, I the capital cost, and V the venture profit. The terminology is discussed in Chapter 4. This particular system was chosen deliberately as a marginal case. The second expression is from Schweyer and May (1962) and the venture profits for both cases are the same at the best guess estimates. The manager concerned indicated that the sales volume would be as follows:

optimistic guess	8200 items/annum
best guess	7000 items/annum
pessimistic guess	5000 items/annum

Compare the effect on the profitability estimate of a capital cost of $830 000 ± 15% and $830 000 ± 10%.

Assuming the estimates fall within the 2σ confidence limits and differentiating the expressions to generate the sensitivity coefficients the variance of the venture profit is calculated in Table 3.3.

Table 3.3. Effect of sales and income variance on profit

	Case 1 Capital cost ±15%	Case 1 Capital cost ±10%	Case 2 Capital cost ±10%	Case 2 Capital cost ±15%
σ_M	800	800	800	800
σ_I	62 250	41 500	62 250	41 500
$S_M = \dfrac{\mathrm{d}V}{\mathrm{d}M}$	23·3	23·3	8·16	8·16
$S_I = \dfrac{\mathrm{d}V}{\mathrm{d}I}$	−0·19	−0·19	−0·19	−0·19
Sales: $S_M{}^2\sigma_M{}^2$	$3{\cdot}47 \times 10^8$	$3{\cdot}47 \times 10^8$	$0{\cdot}45 \times 10^8$	$0{\cdot}45 \times 10^8$
Capital: $S_I{}^2\sigma_I{}^2$	$1{\cdot}4 \times 10^8$	$0{\cdot}62 \times 10^8$	$1{\cdot}4 \times 10^8$	$0{\cdot}62 \times 10^8$
Overall: $\Sigma S_i{}^2\sigma_i{}^2$	$4{\cdot}87 \times 10^8$	$4{\cdot}09 \times 10^8$	$1{\cdot}85 \times 10^8$	$1{\cdot}07 \times 10^8$
$\sigma_u = \sigma_v$	$22\,050	$20\,200	$13\,600	$10\,330

Note that Case 1 represents the more common occurrence when the effect of errors in capital cost of ±15% and ±10% make little difference in decision making compared with sales variance.

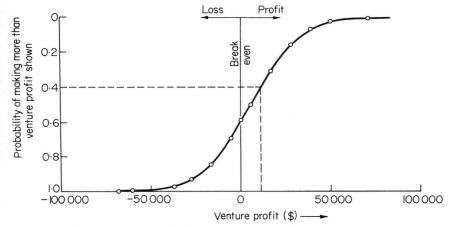

Figure 3.2 Probability of achieving more than venture profit

If further 'uncertainty' analysis (there is nothing as certain as uncertainty) is carried out, the meaning of variance is clarified.

Using the previous data for Case 1 the venture profit for the best guess is given by

$$V_{best} = (50·8 \times 7000 - 12 \times 7000 - 0·15 \times 830\,000)\,0·6 - (830\,000 \times 0·1)$$
$$= \$5260$$

Similarly the maximum and minimum profits are given by

$$V_{max} = \$56\,850; \qquad V_{min} = \$(-64\,970)$$

Thus it is possible to observe that the profit will fall in the range $\$(-64\,970) \rightarrow \$56\,850$.

Now taking $\sigma_v = 22\,050$ from Table 3.3 and $\bar{E} = 2600$ then values of z can be generated from the formulae $z = (E - \bar{E})/\sigma$. Using probability tables Figure 3.2 is generated. From these data it can be shown that there is only a 40% chance of making less than the minimum acceptable return.

Ross (1971) outlines the technique of uncertainty analysis. In order to highlight various aspects we avoided the non-normal distribution of the sales curve in the example. Computer programs are available for assessing other distributions and Monte Carlo methods (Meyer, 1956) may be used to integrate the probability curves for all the variables. They essentially operate by making repeated computations based on random samples of the various probability distributions. Further examples are given by Dagnall (1965) and Tayyabkhan and Richardson (1965).

4 Economic Analysis for Design

4.1 Introduction

Economic analysis is as essential to the design as charting procedures. Simple rules are developed in this chapter. Examples include a study of a typical balance sheet and an economic comparison of plants to show that the correct criteria must be used to assess project viability.

The initial problem for the process engineer is to select from the considerable number of techniques for deciding whether capital expenditure is worth while. A method is required specific to each design problem. Valid values must be substituted in the cost equations. Of particular importance is the minimum acceptable return (that the company expects in order to justify capital expenditure on the project) and incorporating the latest changes of government policy with regard to allowances and taxation. For this reason it is dangerous to use rule of thumb for decisions on the optimum size of equipment, piping insulation, etc.

4.2 Balance sheets for chemical plant

A profit and loss table is indicated Table 4.1. The column indicating variance from standard has been omitted. The factors making up the table are discussed as follows.

(a) *Value of production.* An important factor in the profitability of an installation is the rate at which it operates after it has been built. Despite the apparently favourable aspects of a project, if the outlets have been badly calculated and the plant can operate at no more than, say, 0·7 of capacity, it will have difficulties competing with another plant less advantageously conceived but able to operate at its maximum nominal capacity. Value of production includes any by-products. All material is not necessarily sold at the same price and the value included in the previous table should be an average, computed separately (see also items (j) and (k)).

(b) *Cost of raw materials.* Correct yields are essential. Too often at the authorization stage only the yield of a process from considerations of the reaction is included. Losses occur in various ways: liquid remaining on

20

Table 4.1. Typical profitability summary (before taxes and depreciation)

(1) *Naphtha, steam cracking process*		400 000 t/annum ethylene	
Investment (grass roots plant)		$U.S.	
(a) Process units, battery limit		10 000 000	
(b) Offsites		10 000 000	
(c) Start up expenses and proportion of royalties		4 000 000	
(d) Total investment (a)+(b)+(c)		54 000 000 (*I*)	
(2) *Value of production*		$/annum	$/annum
(a) Ethylene	400 000 t at 100 $/t	40 000 000	
(b) By-product sales:			
Propylene–propane	255 000 t at 30 $/t	7 650 000	
Butylene–butadiene	100 000 t at 30 $/t	3 000 000	
Gasoline	275 000 t at 22 $/t	6 250 000	
Fuel oil	45 000 t at 15 $/t	675 000	
(c) Total value of production (a)+(b)			57 575 000
(3) *Raw materials*			
(a) Naphtha 1 425 000 t at 18 $/t		25 650 000	
(b) Total raw materials			25 650 000
Value of production (less raw materials) (2)–(3)			31 925 000 (*S*)
(5) *Variable operating cost*			
(a) Utilities			
(1) Fuel 180 000 t at 18 $/t		3 240 000	
(2) Steam (credit) 1 000 000 t at 1·6 $/t		1 600 000	
(3) Electricity 420 000 000 kW at 1 C/kW h		4 200 000	
(4) Cooling water 80 000 000 m^3 at 0·6 C/m^3		480 000	
(5) Boiler feed water 2 000 000 m^3 at 0·2 $/m^3$		400 000	
(6) Total utilities (1) to (5)			6 720 000
(b) Catalyst and chemicals			
(1) Dessicant 200 000 kg at 0·3 $/kg		60 000	
(2) Caustic 3 000 t at 70 $/t		210 000	
(3) Catalyst 30 000 kg at 4 $/kg		120 000	
(4) Total catalyst and chemical (1) to (3)			390 000
(c) Labour and supervision			
(1) Salaries/wages, 120 men/annum at $7000		840 000	
(2) Supervision at 25% salaries and wages		210 000	
(3) Total wages (1)+(2)			1 050 000
(d) Total operating cost, variable element (a)+(b)+(c)			8 160 000 (*C*$_v$)
(6) *Expenses*			
(a) Maintenance (4% of investment)		2 000 000	(*m*I)
(b) General expenses (3% of investment)		1 500 000	(*g*I)
(c) General overheads (3% of investment)		1 500 000	(*o*I)
(d) Total expenses (a)+(b)+(c)			5 000 000
(7) *Total operating costs* (5)+(6)			13 160 000 (*C*$_T$)
(8) *Gross return* (before taxes and depreciation)			18 750 000 (*R*)

dumped solid; relief valves leaking; operator's error; excess usage; start-up losses; inventory; etc. These factors can be larger if the conversion per pass through a reactor/recovery system is small, as owing to recycle a considerable amount of feed material must pass round the system for an equivalent amount of product. If 100 kg/s of raw materials were passing through such a system from which 10 kg/s of product was produced per pass, then, if 2% of the total material was lost, the loss related to product would be 2 in 10, or 20%. When considering the purchase and price of the materials the effect of purity on the process must be considered. Adaption of the process should be made if an improvement in profitability is possible by a change of supply (see also section (k)).

(c) *Energy costs.* Always ensure the figures represent true costs and are not based on some convenient low cost from one plant to another to improve the balance of the latter. Costs appertain not onlyto the process but also to warehouses, control-room heating, etc. In addition, hot stand-by and start-up have to be allowed. When ascertaining electrical charges allow for possible changes in the maximum demand of the works.

(d) *Catalyst and chemicals.* Miscellaneous chemicals are often required, particularly acids and alkalis. Always include these costs, although the initial catalyst charge is usually assigned to capital expenditure. Chemical dosing of cooling water is often incorporated in this section rather than under cooling water costs.

(e) *Wages.* Here direct labour and supervision only are included. Estimation of requirements may be by comparison with similar plants, adjusting for the degree of automation. Allowances are made for wages costs and the coverage provided for holidays.

(f) *Maintenance.* For a new project this is usually assessed as a percentage of capital cost. Common values vary from 3 to 10%/annum according to the type of process and main materials of construction. As soon as possible install actual costs which should be subdivided into manpower, materials, and contract work.

(g) *General expenses.* The heading usually includes rates and insurance; laboratories—process control and direct research; site charges such as internal transport and rubbish disposal; some administration: say, 3% of capital per annum for new projects.

(h) *Overhead expenses.* The overhead expenses apply for the directorate; long-term research; commercial department; accountancy; canteen; etc.: say, 4% of capital per annum for new projects. All plants whether new or old must, if at all possible, contribute towards these overheads and general expenses. It may seem unfair to burden a new plant with such expense if this is installed on a fully serviced site, but the next plant might involve a new site with services to be supplied. It could never pay for these by itself and the other plants of the company must contribute. The one exception is for the existing plant which only pays its way on marginal costing (see (j)).

(i) *Net profit.* The ensuing results show the gross profit before taxation and depreciation for a particular period (in this case one year). On a new project the profitability is then calculated by the appropriate procedure. For an existing plant the gross profit and its variation from standard is the criteria.

(j) *Variation of profit with throughput.* Items (a) to (d) largely vary with throughput and are thus variable costs, i.e. as the throughput varies so these costs adjust. Items (e) to (i) are fixed costs and if throughput is low will dominate the profit of the plant. The importance of these is shown by comparing throughputs, Table 4.2. The balance sheet is shown in modified form for ease of calculation, the difference between fixed and variable costs being better defined.

Table 4.2. Effect of throughput on profit

Output (t/annum)	800	900	1000
Selling price (£/t)	50	50	50
Material costs (£/t)	20	20	20
Variable costs (£/t)	5	5	5
Contribution (£/t)	25	25	25
Total contribution (£/annum)	20 000	22 500	25 000
Fixed costs (£/annum)	5 000	5 000	5 000
Depreciation (£/annum)	5 000	5 000	5 000
General overheads and expenses (£/annum)	4 000	4 000	4 000
Total profit before taxes (£/annum)	6 000	8 500	11 000

Incidentally, note that 'direct costs' represent all the costs directly related to the plant (i.e. excluding overheads); 'marginal costs' represent the costs which, if the plant were continuing in operation, would be incurred and allows for all materials and services to be charged into the plant at marginal prices themselves; 'incremental costs' represent the actual cost of producing an additional increment of material.

The large variation of profitability with throughput makes the initial sizing of plant so difficult. If the sales do not reach the target, the profit forecast is greatly reduced.

Throughput on a plant may be reduced by the operating problems of the production department. The planned stoppages which are incorporated in maintenance and operating schedules are easy to assess. More difficult to assess are problems arising through product being off-specification and requiring re-working/blending, or failure of a plant item causing shut-down of the process. Commissioning can be particularly fraught with these problems and, on continuous plants with little storage, chain reactions to other plants can be set up. This means that this important aspect of profitability can be most difficult to forecast.

The chance of simultaneous success of three operations is for the alternative structures of series and parallel (see p. 90):

$$\text{series}: R_s = R_1 R_2 R_3$$
$$\text{parallel}: R_p = 1-(1-R_1)(1-R_2)(1-R_3)$$

where R is the reliability.

Obviously, if for a system the reliability works out at 96 successes for every 100 trials and it feeds another system of reliability 96/100, the combined reliability assuming negligible inter-process storage is 92/100. It is best to use hours lost through failure when assessing success as a shut-down can result in a disproportionate loss in production. Thus the availability of the process is made up of the engineering reliability of the process, the planned stoppages, and operators' reliability. Values depend on the type of plant, age, etc. One general figure is to assume an onstream time of 8000 h/annum. Clearly in the early days of the plant it is wiser to take a much lower figure.

(k) *Transfer prices.* A problem which often arises is at what price should material be transferred between plants. This applies to utilities as well as chemical products. Every case has to be considered separately, but a few rules of thumb may be prove useful.

New project. For major streams charge the plant full costs and base transfer costs on outside markets. If a new project cannot pay its overheads, it is hardly the best of enterprises. Once marginal costing is accepted for such a plant, this procedure can spread and that way lies economic ruin.

For minor costs, say, when optimizing insulation thickness, I would use marginal or incremental costs in the process engineering design calculations, providing adequate capacity exists on the plant which supplies the material being transferred, steam in this case. It represents more accurately the economic picture and the costs can easily be lost in the overall operating costs. Thus on paper the accountants will be happy and a minor investment opportunity has not been lost.

Existing plant. A distinction may be made between the planning model and the plant return. For centralized planning between companies in the group or individual plants it is recommended to use incremental costs if the material could not otherwise be sold outside. If an outside market exists use full values unless this product cannot be purchased. For individual plant returns always use full values.

Constrained plant. If the internal requirement is subject to a constraint such that the increase is going to result in a demand for, say, extra steam capacity or exceeding the maximum demand for electricity on the works, then a scarcity factor must be used in reassessing the entire other users.

4.3 Economic criteria

4.3.1 GENERAL

The process engineer requires a knowledge of economic criteria for two aspects of his work and the same criteria need not be used for both. They are (A) to make a decision with regard to the overall viability of the process he is considering; (B) to enable him to carry out economic balances in the process design.

A criterion is required which prevents waste of time on abortive applications for capital expenditure and a rough approximation such as pay-back

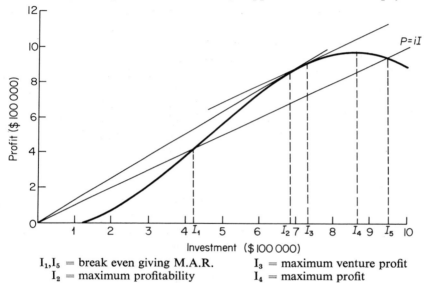

I_1, I_5 = break even giving M.A.R. I_3 = maximum venture profit
I_2 = maximum profitability I_4 = maximum profit

Figure 4.1 Alternative criteria for deciding an investment

time may suffice. However, often the process engineer is involved in assisting in the preparation of financial data for the application for capital expenditure, so he requires a knowledge of the mechanism of the economic criteria being used. Profit may be expressed as a balance sheet, in algebraic form, or as a graph, Figure 4.1 (the latter enables the range of possible selections to be clearly seen). The minimum acceptable return concept derived by Happel (1958) states that before an investment should proceed it is necessary for a process to give a minimum rate of interest. If this return is deducted from the overall profit the curve of venture profit is generated. Which of the two major alternative best investments I_2 or I_3 is selected depends on the company policy on a specific project. Adopting algebraic nomenclature we are saying that for an investment to proceed

$$\left(\frac{P}{I}\right)_{e.c.} > i_b$$

where subscript e.c. modifies profit and investment according to economic criteria used and interest b refers to the base case of the project. The aim of the project may be to optimize, say, either the short-term profitability

$$\max \left(\frac{P}{I}\right)_{\text{e.c.}}$$

or the overall cash flow

$$\max \left(P_{\text{e.c.}} - i_{\text{b}} I_{\text{e.c.}}\right)$$

An alternative approach is required when using an economic criterion in conjunction with the process design. Here the process engineer should use an amenable algebraic expression. It does not matter whether this is a method approved by the companies' finance department because such a criterion is an integral part of the engineers' design equations. Happel (1958) has pioneered much of this work and Table 4.3 indicates some of his most used simple algebraic expressions and their derivation. The expressions may be used to balance operating costs and capital costs in design calculations such as distillation, pipework, thickness of insulation, etc., where an optimum value is produced. In order to identify the algebraic values these have also been indicated on the balance sheet, Table 4.1.

4.3.2 TIME VALUE OF MONEY

The expressions are valid for some calculations but ignore the time value of money. Future earnings are preferably related back to their present value.

If $S(\theta)$ is the amount in a fund in time θ and interest i be continuously compounded, in the increment of time $d\theta$, the change in the amount of the fund is

$$dS = iS \, d\theta$$

Therefore the amount of money at $\theta = 0$, $S(0)$ accumulating to $S(\theta)$ at time θ, is obtained by integrating from $\theta = 0$ to θ (*Computers in Engineering Design*, 1966).

$$S(0) = S(\theta) \, e^{-i\theta}$$

Alternatively one can consider a sum of money at $S(\theta)$ which earns interest i each year. Then the money grows as follows:

$$S(0)(1+i); \qquad S(0)(1+i)^2; \qquad S(0)(1+i)^3$$

and, if $S(\theta)$ is brought back to the present time, *the discounted annual return* (Merrett and Sykes, 1965) is

$$S(0) = \frac{S(\theta)}{(1+i)^\theta}$$

The two approaches give slightly different answers and continuous compounding of money is a theoretical concept which gives a more amenable

Table 4.3. Simple venture profit

---

(1) *Derivation of venture profit*
Capital investment $= I$
Total sales realization $= S$
Total running operating costs
$$C_T = C_v + mI + rI + oI + gI$$
where C_v = variable costs + wages
$\quad m$ = % maintenance costs
$\quad o$ = % overhead costs
$\quad p$ = % royalties
$\quad g$ = % general expenses
Gross return $= R = S - C_T$
If depreciation $= dI$
Taxes $= t$
Profit $= P = (R - dI)(1 - t)$
Venture profit $= V = P - i_m I$
where i_m = minimum acceptable return (M.A.R.)

---

(2) *Definition of U*
U is an algebraic term which equals the total costs before taxes, *note* including M.A.R. term.
By definition
$$V = (S - U)(1 - t)$$
i.e. $\quad V = P - i_m I = (R - dI)(1 - t) - i_m I$
$\quad\quad\quad = (S - C_T - dI)(1 - t) - i_m I$
$\quad\quad\quad = (S - C - mI - oI - gI - dI)(1 - t) - i_m I$
So for straight-line depreciation
$$U = C + mI + oI + gI + dI + i_m I/(1 - t)$$

---

(3) *For additional equipment*
$$V_1 = P_1 - i_m I_1$$
$$V_2 = P_2 - i_m I_2$$
$$\Delta V = \Delta P - i_m \Delta I$$
if $\Delta V > 0$; $\Delta P/I > i_m$

---

algebraic expression. For process engineering the future earnings used in design equations should be additive. For instance, when designing a waste heat boiler and considering the incremental returns of the saving from the last square foot of surface area it is desirable to subtract one profit from the other. Thus an additive economic criterion should be used, i.e.

$$\text{N.P.V. (a)} - \text{N.P.V. (b)} = \text{N.P.V. (a} - \text{b)}$$

where N.P.V. is net present value. Because of cancellation of terms several criteria can be used in this way, but notably not discounted cash flow. Note that for an incremental investment the marginal case is when

$$\left(\frac{P}{I}\right)_{e.c.} = i_b$$

4.3.3 SELECTION OF INVESTMENT CRITERIA

An investment criteria should accurately determine whether to proceed with the project. This must incorporate time value of money, permit a high level of judgement, and accurately establish the risk to capital involved. At the same time it should reflect ordinary management attitudes and not require 'a bank manager in the cupboard' to understand. The degree of complexity should be capable of expansion with the availability of data.

For the process engineer especially it should be capable of algebraic manipulation and incorporation within the engineering design. If the objective function as reflected by profitability is, say, to maximize short-term earnings or to maximize the rate of earnings on products, then the economic criteria must also adapt. For surveys of investment criteria see Abrams (1970), Perry (1963), and Rudd and Watson (1968). The conclusions from these surveys supported two systems.

Net present value N.P.V.: the sum of cash flows discounted to time zero, taken as the year of the first outlay.

$$\text{N.P.V.} \sum_{\theta=0}^{\theta=n} \frac{S_{(\theta)}}{(1+i)^\theta}$$

Venture worth W: a minimum acceptable return on capital i_m is determined in addition to the cost of capital i. Capital is divided into the risk free, I_w (working capital, services), and at risk, I_p (the capital generated by the project). If the venture profit is defined as

$$V = P - I_w - i_m I_p$$

a simplified venture worth is

$$W = V^{-i\theta} \, d\theta$$

The expanded values for the equations are more complex, for example, see the venture worth equations, Table 4.4 (Happel, 1958). However, given the necessary data no mathematical problems arise.

Abrams favoured N.P.V. and Rudd and Watson recommended venture worth. Abrams' main objection to venture worth was expressing risk by a function of the rate of interest used. Rudd preferred venture worth as the minimum rate of return was a useful design criterion.

Let us consider some examples to see how economic criteria may be used.

Example 4.1. Alternative criteria for investment. Selection according to

Table 4.4. Financial equations

Financial equations may be developed the same as for any unit operation. These equations are typical of the venture profit/worth system (Happel, 1958) (assuming sinking fund and depreciations same).

Simple venture profit

$$V = P - i_m I = (R - dI)(1 - t) - i_m I$$

Venture profit

$$V_K = R_K - \frac{iI}{(1+i)^n - 1} - (R_K - d_K I)t - i_m I - i I_w - \frac{I(1-t)Sa}{(1+i)^n - 1}$$

| | minimum acceptable return | working capital | salvage |

which may be modified to venture worth W as shown.

$$W = \sum_{k=1}^{k=N} \frac{V_k}{(1+i)^k} = \sum_{k=1}^{k=N} \frac{(1-t)R_k}{(1+i)^k} + \sum_{k=1}^{k=r} \frac{d_k tI}{(1+i)^k} - \left\{ \frac{(1-i)^N - 1}{i(1+i)^N} \right\} (i_m - i)(I + I_w)$$

| | summation of gross untaxed income | summation of discounted tax credit | present worth of incremental return on total investment |

$$- I - \left\{ \frac{(1+i)^N - 1}{(1+i)^N} \right\} I_w + \frac{(1-t)Sa}{(1+i)^n}$$

| | initial investment | present worth of cost of working capital | present worth of salvage |

the criteria indicated in Figure 4.1 has been analysed using the expression in Example 3.1 and the following procedure outlined by Schweyer and May (1962).

$$P = (76M - 0.0036M^2 - 12M - 0.15I)(1 - t)$$

If we assume the investment is given by

$$I = 4000M^{0.6}$$

where M is the production rate/annum, P is the profit after taxes (t), and I is the investment. Then

$$P = \left\{ 64 \left(\frac{I}{4000} \right)^{1.67} - 0.0036 \left(\frac{I}{4000} \right)^{3.33} - 0.15I \right\}(0.6)$$

or

$$P = 3.705 \times 10^{-5} \times I^{1.67} - 2.185 \times 10^{-15} \times I^{3.33} - 0.09I$$

Maximum profit is given when the derivative of P with respect to I is equated to zero

$$\frac{dP}{dI} = 6.187 \times 10^{-5} \times I^{0.67} - 7.276 \times 10^{-15} \times I^{2.33} - 0.09 = 0$$

which maximizes at $I_4 = \$868\,000$.

Assuming M.A.R. of $0.1I$, then maximum venture profit is obtained from

$$\frac{dV}{dI} = 6{\cdot}187 \times 10^{-5} \times I^{0{\cdot}67} - 7{\cdot}276 \times 10^{-15} \times I^{2{\cdot}33} - 0{\cdot}19 = 0$$

which maximizes at $I_3 = \$732\,000$.

Profitability is given by

$$100\frac{P}{I} = 3{\cdot}705 \times 10^{-3} \times I^{0{\cdot}67} - 2{\cdot}185 \times 10^{-13} \times I^{2{\cdot}33} - 9$$

which maximizes at $I_2 = \$670\,000$ giving a profitability of $12{\cdot}47\%$.

The results are shown in Figure 4.1 which has been discussed earlier.

Example 4.2. Time value of money. It is certain that £100 today could be invested at 5% rate of interest and be worth £105 in one year's time. Thus, conversely, £105 in one year's time is worth £100 at present value. A discount factor can be used to calculate what the present value of a payment of £1000 payable in six years' time at 15% rate of interest, i.e.

$$1000 \times \frac{1}{(1{\cdot}15)^6} = £432$$

Consider a project for £5000 where the earnings are as follows:

1st year	£2000
2nd year	£1000
3rd year	£1000
4th year	£2000
5th year	£500

Table 4.5. Present value

Year	Cash flow (£)	Discount factor	Present value (£)
0	−5000	1·0000	−5000
1	+2000	0·9091	+1818
2	+1000	0·8265	+ 826
3	+1000	0·7513	+ 751
4	+2000	0·6830	+1366
5	·+ 500	0·6209	+ 310
5	+1000	0·6209	+ 621
			£+ 629

At the end of five years the salvage value is £1000. Using tables for obtaining the discount factors, the net present value at 10% rate of interest may be calculated as in Table 4.5.

The net present value of this project is £629. The project will earn 10% on the capital invested over its life and an additional sum equivalent to £629 today.

The discounted cash flow (D.C.F.) rate of return is calculated by trial and error. It is defined as the rate of return, i, which gives a zero net present value (Merrett and Sykes, 1965), 12% in this example.

Example 4.3. Cash flows in an actual project. Table 4.6 shows how the cash flows on a simple system might build up. Note how taxation is not paid in the same year as the earnings rise. A decision whether to go ahead with the project depends if these cash flows are adequate. Depreciation has been calculated by declining balance.

Example 4.4. Economic criteria have no meaning after investment. A feature of the economics of investment criteria is, once the expenditure is made and the plant installed, they no longer have any meaning.

To illustrate this point consider a plant which can be increased in output by a series of staged expansions. Information is available on the present plant and the capital then expended. The plant at present is running below earlier forecasts and giving a return on the original scheme of 11% after taxes. An acceptable return on this project would be 15% after taxes. Income taxes are 50% and depreciation permitted as a straight line over 10 years. Analysis is as follows.

	Base	Case 1	Case 2	Case 3	Case 4
Cumulative capital I (£)	21 000	24 400	29 000	32 000	35 200
Total gross annual profit R (£)	6 820	8 920	10 940	11 320	12 800
Depreciation dI (£/annum)	2 100	2 440	2 900	3 200	3 520
Gross profit before taxes (£/annum)	4 720	6 480	8 040	8 120	9 280
Profit after taxes P (£/annum)	2 360	3 240	4 020	4 060	4 640
Profitability P/I (%)	11	13	14	13	13

The above profitability is meaningless as, once the capital is spent, providing the plant is making a profit under marginal costs it should remain in operation. Only incremental costs can be compared. These can be obtained by subtracting each case from that preceding.

Differential profit P (£/annum)	880	780	40	580
Differential capital I (£/annum)	3400	4600	3000	3200
Profitability P/I (%)	26	17	1·3	18

Table 4.6. Simple cash flow

Year	Capital expenditure (£m)	Production (ton)	Sales revenue (£m)	Operating costs (£m)	Operating margin (£m)	Taxation allowance at 15% × 50% (£m)	Taxation excluding allowance at 50% (£m)	Net cash flow (£m)	N.P.V. (£m)	Cumulative N.P.V. referred to year 0 (£m)
1	4	0	0	0	0	0	0	−4·00	−3·636	−3·363
2	4	0	0	0	0	0	0	−4·00	−3·306	−6·942
3		30 000	3	2·9	0·1	0	0	+0·10	+0·075	−6·867
4		60 000	6	3·6	2·4	0·60	−0·05	+2·95	+2·015	−4·852
5		65 000	6·5	3·8	2·7	0·51	−1·20	+2·01	1·248	−3·604
6		70 000	7	4	3	0·43	−1·35	+2·08	1·174	−2·430
7		70 000	7	4	3	0·37	−1·50	+1·87	0·960	−1·470
8		70 000	7	4	3	0·31	−1·50	+1·81	0·845	−0·625
9		70 000	7	4	3	0·27	−1·50	+1·77	0·751	+0·126
10		70 000	7	4	3	0·23	−1·50	+1·73	0·667	+0·893

(i) N.P.V. discounted at 10%.
(ii) Discounted cash flow rate of return 12% after 10 years.
(iii) Working capital omitted.

Thus carry out schemes 1 and 2. Reject scheme 3. As scheme 4 cannot be carried out unless 3 is completed, combine 3 and 4, and compare with 2. For 3 and 4, the differential return is 10%, so reject. Recommendation: expand to scheme 2 and no further.

Example 4.5. Pipeline sizing. Pipework costs can be 20% of the overall capital costs of the plant, so it is important to be accurate. The problem is to balance operating and capital charges. As the pipe diameter gets smaller, capital charges decrease and pumping costs increase.

The basic formulae of Generaux (1937) is still applied, although, in general, nomographs (Perry, 1963) and rules of thumb are used which assume values in this equation. Applying venture profit (Happel, 1958) and allowing values for allowances, grants, etc., a formula has been devised which permits upgrading of this information (Bush and Wells, 1971).

Typical results are listed in Table 4.7 for varying some parameters. They are obtained by the substitution of discrete pipe sizes into the design equations.

Table 4.7. Optimum pipe diameters

Flow (ft^3/min)	150	150	150	150	150	150
Maintenance (%I)	0·06	0·06	0·06	0·06	0·06	0·06
Minimum acceptable return (%I)	0·5	0·5	0·5	0·5	0·5	0·1
Grant (%I)	0·5	0·5	0·5	0·5	0	0
Head (ft)	1800	1800	1800	1800	100	100
Electricity (p/kW h)	2	1	2	2	2	2
Actual minimum pipe size (in)	8	7	7	9	7	9
Actual pipe selection (in)	8	6	6	10	8	8

A value taken from the nomograph ignoring the change in the economic information is 10 inches. Increase in minimum acceptable return and maintenance costs reduces the size; whilst increase in other factors increases the optimum. Clearly allowing for these factors has a greater effect at large diameters of pipe and for 3 inch diameter minimum size pipe the nomograph may be misused by the user as sufficient accuracy is possible. Results for stainless steel pipe show smaller sizes than noted above.

Example 4.6. Pump Selection. Happel's (1958) expression

$$U = C + mI + oI + gI + dI + i_m I/(1-t)$$

is useful for rapid evaluation of alternative items. It represents the case when sales are unaffected by the process decision and operating costs and charges on capital are to be minimized.

Consider the selection between two pumps A and B which are equally reliable.

$$\text{Pump } A: 84\% \text{ efficient, 30 kW, } I = \text{£3600}$$
$$\text{Pump } B: 90\% \text{ efficient, 30 kW, } I = \text{£4000}$$

Depreciation 0·1; power 1 p/kW h; taxes 0·4; maintenance 0·07; 8000 h/annum; M.A.R. 0·12.

$$U_A = \frac{30}{0\cdot84} \times 0\cdot01 \times 8000 + 3600\left(0\cdot07 + 0\cdot10 + \frac{0\cdot12}{0\cdot6}\right) = \text{£4189}$$

$$U_B = \frac{30}{0\cdot9} \times 0\cdot01 \times 8000 + 4000\left(0\cdot07 + 0\cdot10 + \frac{0\cdot12}{6}\right) = \text{£4147}$$

Thus select pump B.

If the pumps were purchased with a spare and for simplicity, assuming maintenance costs not increased,

$$U_A = 2857 + 7200\,(0\cdot035 + 0\cdot1 + 0\cdot2) = \text{£5269}$$
$$U_B = 2667 + 8000\,(0\cdot035 + 0\cdot1 + 0\cdot2) = \text{£5347}$$

Thus select pump A.

Example 4.7. A bid-tabulation. The analysis of a bid-tabulation was mentioned in Chapter 3. Let us expand on this by an example. Table 4.8 shows a typical bid-tabulation for some compressors. Which would you select?

Let us put forward some ideas.

Any assumptions made are not necessarily accurate and merely indicate lines of argument. Other reasons could no doubt be produced to change such arguments.

Note. Care should be taken that in carrying out such analysis there is no danger in affecting previous optimization conclusions. Do not carry out improper sub-optimization of an interior component (Bellman, 1957) (see dynamic programming).

Economic basis. For this simple exercise it is only necessary to balance increased operating costs with factors which increase with capital. Of the latter only maintenance, depreciation, and acceptable return on capital need be considered. A modified venture profit has been used as indicated in Table 4.3.

Factors

(a) Direct drive is assumed to give 85% overall power efficiency. Other drives are assumed at 80% overall power efficiency.

(b) Cast iron cylinder increases maintenance costs by 4%I.

(c) Flat belt drive increases maintenance costs by 5%I.

(d) Horizontal machine preferred to vertical but this is balanced by longer stroke and increased foundation costs.

Description	Special requirements	Manf'r 1	2	3
(1) model and size duty	all models are the correct duty	TM 211S	TM 211D	AB75
(2) No. stages; No. cylinders; single/double		1; 1; SA	1; 1; DA	1; 1; SA
(3) design		vertical V-belt	vertical V-belt	horizontal V-belt
(4) cylinder diameter; piston stroke		4; 5	4; 5	H 3½; 7
(5) rev/min; Bhp of compressor		150; 12	500; 10	413; 12·5
(6) materials		C.I. liner	C.I. liner	C.I. cylinder
(7) auxiliaries		supplied	supplied	supplied
(8) price (£)		2650	3020	2280
(9) driver (hp); cost (£)		14; 1460	14; 1460	14; 1460
(10) extras		—	—	—
(11) delivery months		8–10	9–10	10
(12) total cost (£); incl. spare hp		8220; 12	8960; 10	7480; 12·5

Description	Manf'r 4	5	6	7
(1) model and size duty	ACE 3	HB 3	AQ 108	US 10
(2) No. stages; No. cylinders; single/double	1; 1; DA	1; 1; DA	1; 1; DA	1; 1; DA
(3) design	vertical V-belt	Horizontal V-belt	vertical direct	vertical flat belt
(4) cylinder diameter; piston stroke	4; 6	3; 7	3; 5	4; 4
(5) rev/min; hp of compressor	500; 12	4·65; 12	735; 12	400; 11
(6) materials	C.I. liner	C.I. liner	C.I. liner	C.I. liner
(7) auxiliaries	supplied	supplied	supplied	supplied
(8) price (£)	4170	2620	2865	2130
(9) driver (hp); cost (£)	16; 1600	15; 1460	15; 1460	13·5; 1460
(10) extras	£800 duty	£750 duty	—	—
(11) delivery months	14	7	8–10	11
(12) total cost (£); incl. spare hp	13 140; 12	9660; 12	8650; 12	7180; 11

Remarks. For this plant $i_{M.A.R.}$ = 0·15; taxes = 0·5; depreciation = 0.1. Constant maximum production rate. Delivery not critical except for item (4).

(e) To reduce the detail it is assumed that other installation costs were the same for each unit.

Evaluation: i_m, 15%; taxes, 50%; depreciation, 10%. Maintenance: 4%.

Charges on capital $= \dfrac{0\cdot15}{1-0\cdot5} + 0\cdot1 + 0\cdot04I = 0\cdot44I.$

Power charge at $0\cdot01$ p/kW h; hours on stream $= 8000$.

One Bhp difference $\equiv \dfrac{1}{0\cdot8} \times 0\cdot75 \times 8000 \times 0\cdot01 = £40$

Comparison. Work down the list selecting best buy each time.

Machines	Differential operating costs		Selection
1 and 2	$\Delta I = 740$; charges 326	$> 2 \times 40$	1
1 and 4 and 5	1 superior in both respects		1
1 and 6	$\Delta I = 430$; charges 190	$> 12 \times 40 \times 0\cdot05/80$	1
3 and 1	$\Delta U = \quad 8220 \times 0\cdot44$	$> 0\cdot5 \times 40$	marginal
	$-7480 \times 0\cdot48 = 31$		3*
7 and 3	$\Delta U = \quad 7480 \times 0\cdot48$		
	$-7180 \times 0\cdot53 = -219$	$< 1\cdot5 \times 40$	3

Decision. Select compressor from manufacturer 3.

*Marginal decision. 3 selected as felt that 4% surcharge on maintenance was if anything high.

4.3.4 DISCUSSION OF THE EXAMPLES

(i) *Overall profitability.* Selection as to which method to use for assessing overall profitability is rarely the function of the process engineer and quite correctly depends on a decision within the company as to which criteria they can best interpret from their accumulated experience (for a review of investment appraisal, see Betts (1972)). Clearly it is best that this incorporates time value of money (Example 4.3) but, as economic criteria have no meaning following investment (Example 4.4), mathematical exactness is not required. As long as the decisions are correct the company will continue in sound financial health. Of course it can be argued they might do better if their interpretation was more refined (Example 4.1) and doubtless uncertainty analysis (Example 3.1) will become wider used. This will improve the lot of the process engineer if it succeeds in removing the pressure of achieving accurate capital cost estimation by focusing attention on other factors which affect the profitability estimate to a greater extent.

To amplify Example 4.4, the policy decision on a new venture depends on the management being reasonably assured that a required throughput will be achieved and the selling price of the products will cover all expenses, will make a contribution to overheads, and will produce the required rate of profit on the capital employed. Once the money is spent the management are considering the operating profit of an existing plant and considerable reductions in selling price can be made before the company is better off to discontinue production on that plant. Whether or not 'marginal' sales business should be accepted is a difficult decision. Often news of the sales price cut to one customer reaches others and such a price becomes the standard. Similarly, marginal business might preclude later business elsewhere at regular prices. Price wars with rivals may ensue.

(ii) *Economic design criteria.* The decision of which economic criteria to use in a process calculation is made by the process engineer. It depends on the specific problem and availability of data. The examples given of process calculations did not allow for time value of money. This could have been incorporated but it is doubtful whether any decisions would change. If every pipeline were designed in such a manner the design costs would be exorbitant. It is better to average out the plant capacity and work on this basis. Often an excuse for oversizing pipework is that the plant will later increase capacity. This should not be an acceptable reason for poor design and, if expansion is planned, an appropriate percentage should be allowed as a specific decision by the management.

Consider some actual cases. Although pipework represents a considerable percentage of the capital cost of any plant, no accurate economic balance is made. Unfortunately neither client nor contractor make available data on allowances and minimum acceptable return. This is an area in which there is scope for improvement. Similar conclusions were obtained from a study of optimum thickness of insulation (Wells, 1971). The results are indicated in Table 4.9 and show how, if marginal or actual costing is used or a development grant (g) is available, the recommended insulation thickness for a vessel varies. Values from Lyle (1947) and rules of thumb are compared. Without using accurate economic criteria errors are introduced (but see also Section 4.4). Although for insulation this does not represent a major loss for the company, the same decisions arise time and again over equipment. When companies are seriously concerned about their earning value being, say, 8·5% on capital instead of 10%, then the magnitude of these errors is seen.

For pump selection the importance of basing selection on profitability is emphasized. The use of economic design criteria for bid-tabulations is contrary to present practice and arises partly owing to the different objectives of contractors and manufacturing companies. It is very much up to the latter to enforce. The techniques exist and the decision on economic grounds is extremely simple. Such considerations should be supplemented with the experience of the engineer who must assess whether the unit will give reliable service in a specific duty.

Table 4.9. Recommended insulation thickness for vessels

Low heat value $(c = 0.2 \times 10^{-6})$	Thickness (in)				
Internal temperature (°F)	400	600	800	1000	1200
$g = 0$:					
$\quad i_m = 0.15$	$1\frac{1}{2}$	$1\frac{1}{2}$	$2\frac{1}{2}$	$2\frac{1}{2}$	3
$\quad i_m = 0.3$	$1\frac{1}{2}$	$1\frac{1}{2}$	2	2	2
$g = 0.4$:					
$\quad i_m = 0.15$	2	$2\frac{1}{2}$	3	$3\frac{1}{2}$	4
$\quad i_m = 0.3$	$1\frac{1}{2}$	2	$2\frac{1}{2}$	$2\frac{1}{2}$	3

High heat value $(c = 0.6 \times 10^{-6})$	Thickness (in)				
Internal temperature (°F)	400	600	800	1000	1200
$g = 0$:					
$\quad i_m = 0.15$	$2\frac{1}{2}$	$3\frac{1}{2}$	4	4	4
$\quad i_m = 0.3$	2	$2\frac{1}{2}$	3	$3\frac{1}{2}$	$3\frac{1}{2}$
$g = 0.4$:					
$\quad i_m = 0.15$	$3\frac{1}{2}$	4	4	4	4
$\quad i_m = 0.3$	$2\frac{1}{2}$	$3\frac{1}{2}$	4	4	4

Compare from rules of thumb and Lyle (1947)

Internal temperature (°F)	300	500	700	900	1100
Recommended thickness (in)	$1\frac{1}{2}$ ($2\frac{1}{2}$)	2 ($3\frac{1}{2}$)	$2\frac{1}{2}$ (4)	3	$3\frac{1}{2}$

4.4 Economic data for the engineer

It will be noted in the above examples that the engineer requires economic data. These include minimum acceptable return, taxation, and allowances.

What is a minimum acceptable return? To decide this we must return to the authorization stages for capital expenditure. The decision whether or not to go ahead on the project has been made at stage 2, 'budget'. It is most unlikely that the project would be cancelled at stage 3. The problem is whether the minimum acceptable return used to determine authorization at stage 2 should be used for design purposes at stage 3.

Consider the following logic. Rudd and Watson (1968) analysed the corrective factor i_m by subdivision into the following premiums to cover risk:

$$i_m = i + h = i + h_s + h_c + h_I + h_n + h_f$$

where the subscripts refer to sales income (s), manufacturing costs (c), investment (I), project life (n), and probability of complete failure (f)· Their calculated values were high compared with practice but the logic is correct. The items listed above all contribute to an overall assessment of the risk.

Let us assume that at stage 2, budget, the acceptable return was $i+h$. At stage 3 we are studying the case for a proposed heat exchanger which will save steam and cooling water on the plant. If h_s is made up of a proportion h'_s for sales throughput and h''_s for selling price, then only the first factor affects the savings by the exchanger. Thus a theoretical value of $(i+h)_{\text{exchanger}}$ could be built up which was less than $(i+h)_{\text{project}}$, or

$$(i+h)_{\text{project}} > (i+h)_{\text{exchanger}} > i$$

where i is the current earnings rate of the firm.

Further support comes from a study of the exchanger as a separate project. As investment criteria have no meaning once a project is going ahead it might be said that much of the risk on the project has been accepted at the stage 2 approval of capital expenditure. This means that for the heat exchanger we might be considering a different project, one of 'cost reduction on an existing plant'. Risk factors for such a project are much lower as shown by the following values (Happel, 1958).

Type of project	$h = i_m - i$ (%)
high risk—considerably novel project	20–100
fair risk—novel project to company	10– 20
average risk	5– 10
good risk—expansion	1– 5
excellent risk—cost reduction	0– 1

Although such logic is realistic, if money is always funded for acceptable projects it rather upsets the accuracy of the budget estimates. These are required to an accuracy of $\pm 10\%$. Obviously, if the 'budget' design were carried out balancing operating and capital costs for a plant with an i_m of 100%, and then at stage 3 these were altered to an i_m of 15%, this would make nonsense of the estimates.

How can these conflicting factors be balanced. It is recommended at stage 2 that the go-ahead be based on the minimum acceptable return $i+h$ of the project and the engineering design balancing operating and capital costs be based on a lower value $i+h'$, where

$$i+h > i+h' > i$$

For many cases h' will be very small and $i_{\text{M.A.R.}} \approx i$.

In fact this agrees with the present practice. It is unusual for anything other than 'rule of thumb' economic balances to be used. Thus piping nomographs use values i rather than i_m. It is unlikely that this might price some projects out of the market. As shown in Section 3.2, most projects are more sensitive to sales revenue than any other costs.

Of course, if an alternative economic decision is made to minimize capital at risk, then values of $i_{M.A.R.}$ or higher may be used.

4.5 Tax and capital allowances

Taxes and capital allowances must be allowed for in economic evaluations. Legislation changes from year to year, so any listing is indicative only of the assessments. Provision can be made for the following.

(a) Corporation tax at 40% on profits, normally payable 1 to 2 years after the profit is made.

(b) Investment grants which varied from 40% in development areas and 20% elsewhere. Payments were 1 to 2 years after the investment. They were non-taxable and allowances were allowed only on the capital cost less grant.

(c) Capital taxation allowances include the initial allowance which permits anything up to 100% of the asset to be depreciated for tax purposes in the first year. The annual allowance is based on the capital cost less grant and initial allowance usually follows a declining balance.

5 Optimization of Small Systems

Most design procedures for vessels, columns, have an element of optimization built into them. This is often based to a large extent on heuristic methods rather than rigorous, mathematical proof. Now with techniques and economic design criteria becoming available comes the realization that costs must be built into the process engineering procedures.

Let us indicate how the economic design criteria can be incorporated in the optimization of small systems noting that this only results in an optimized subsystem for a specific input from a larger system. Only two procedures are indicated, the general algebraic approach and logical search. For a complete survey see Beveridge and Schechter (1970). The techniques of dynamic and linear programming which assist the optimization of the larger system are discussed in Chapters 11 and 12.

5.1 General algebraic approach

5.1.1 DEFINING THE OBJECTIVE FUNCTION

For a system defined by M variables and N equations the free variables of the system $(M-N)$ require that only $(M-N)$ of the independent variables determining the objective function (Y) should be manipulated. Furthermore it is usually convenient to adjust the system variables without attempting to distinguish between the types of variable encountered, i.e.

$$Y = Y(x_1, x_2, ..., x_{(M-N)}) = Y(x)$$

5.1.2 DEFINING CONSTRAINTS

External agencies such as management and existing utility facilities may exercise constraints on the optimization model. Thus the production rate R may be limited by a maximum capacity C_M:

$$R \leqslant C_M$$

where R is a variable and C_M fixed. Thus

$$R(x_1, x_2, ..., x_n) \leqslant C_M$$

or expressed in the preferred algebraic form

$$R(x_1, x_2, ..., x_n) - C_M \leqslant 0$$

5.1.3 FUNCTION CONTINUITY

The deterministic objective function y, where $y = y(x)$, implies a definite value of the function for any given set of variables x. This function can be continuous, discontinuous, or discrete in form.

It is termed continuous if at some point x

$$y(x) = \lim_{h \to 0} y(x+h)$$

for all the possible ways h can approach zero.

The most common form of discontinuity in a function is the jump discontinuity, e.g. the instantaneous removal of product from a storage tank which is being continuously renewed by production. Another form is found when the function is only valid at discrete values or sets of values of the variables, e.g. pressure drop through pipes where pipes are only available in standard diameters. An optimum may be found under three possible conditions.

(a) In the interior region at a stationary point where all the continuous first derivatives are simultaneously zero.

(b) In the interior region at discontinuities of one or more first derivatives.

(c) At the boundary to the region.

The first case is readily evaluated by setting the partial derivatives equal to zero and checking the stationary values to ensure that these points are not saddle points. Extreme values may also be found at points in the interior of the region at which the partial derivatives are discontinuous. Here again, discontinuities need not necessarily provide extreme values and they are often handled by placing artificial boundaries at them and treating the system in parts. Along boundaries incorporate the boundary restrictions in some modified objective function and use an analytical or search method, a particular example being linear programming. With discontinuous functions, divide the search region by substituting additional boundaries at the discontinuities and use the above methods for handling the continuous sections of the function. For discrete functions it is necessary to compare each value at every point for which the function exists.

5.1.4 SIMPLE ANALYTICAL APPROACH FOR FUNCTIONS OF A SINGLE VARIABLE

As an example of the simple analytical approach, consider the dimensions of a pressure vessel. It is assumed that the capacity of the vessel has been calculated.

Two approaches then apply; either go out to tender, or evaluate the geometry of the vessel before going out to tender. The latter action should permit a fabricator sufficient flexibility that he may select the nearest standard vessel which fulfils the specification should this prove to be cheaper than a one-off special size. It is this approach which will be discussed.

Example 5.1. Determine the optimum dimensions of a closed cylindrical tank which is of volume V, where the costs of the sides is C_s £/unit area and of the ends is C_e £/unit area (Rudd and Watson, 1968; Beveridge and Schechter, 1970).

The total cost equation can be written

$$C_T = 2\pi R^2 C_e + 2\pi R L C_s$$

with $V = \pi R^2 L$, R the radius, and L the length; and on eliminating L then

$$C_T = 2\pi \left(R^2 C_e + \frac{V}{\pi R} C_s \right)$$

Assuming R is unrestricted and C_T is continuous then equating dC_T/dR to zero gives

$$R = \left(\frac{V C_s}{2\pi C_e} \right)^{1/3}$$

and substituting for V in terms of R and L and changing to diameter, D gives the following expression for the minimum total cost (i.e. optimum) diameter/length ratio:

$$D/L = C_s/C_e$$

Thus, if the ends cost three times the sides, then as an approximation the ratio of $D:L$ should be $1:3$.

Example 5.2. The function C_T is not continuous owing to the discontinuous nature of C_e and C_s. Amundson *et al.* (1960) indicate a more precise solution where the discontinuities of C_e and C_s can be accounted for by assuming that certain values apply only within specific regions of the range of the function. In this way the values of R and L at the stationary point corresponding to a pair of values for C_e and C_s can be computed. The R and L at the stationary point might not, however, be values of R and L corresponding to the assumed costs, i.e. within the valid range defined by the particular region of the function, and, if this is the case, we shall say that the stationary point falls outside the permitted region. In other words this particular region does not possess a stationary point.

Reference to Table 5.1 indicates the calculations for a particular volume, $V = \pi/2$. Shown in the final columns are the values of R and L, which yield a stationary point corresponding to the values of C_e and C_s appearing in the

Table 5.1. Stationary values in different regions

Region (see Figure 5.5)	Restrictions on valid range of		Relative costs of plate		Possible stationary values $V = \pi/2$	
	R	L	C_e	C_s	R	$L = \frac{1}{2}R$
1	$\leqslant 1$	$\leqslant 1$	1	1	0·63	1·26*
2	> 1	> 1	2	1·5	0·57*	
3	< 1	< 1	1	1·5	0·72	0·96

* Not valid values.

same line of the table. For $V = \pi/2$ no permitted value exists and so the optimum must be at one or both of the artificial boundaries corresponding to $R = 1$ and $R = 0·71$, i.e. $L = 1·0$ from $L = 1/2R^2$. Costs at the boundaries between regions 3 and 1 and 1 and 2 are indicated below.

Region	L	R	C_e	C_s	C_T
		For $V = \pi/2$			
3	$1·00^+$	$0·71^-$	1	1·5	1·56
1	$1·00^-$	$0·71^+$	1	1	1·21
1	$1·50^+$	$1·00^-$	1	1	1·50
2	$0·50^-$	$1·00^+$	2	1·5	2·75

The signs following numbers are required to indicate direction of approach to discontinuity. For example, 1^+ denotes a number slightly larger than 1. Examining the table we see that the global minimum total cost for $V = \pi/2$ is $C_T = 1·21$, and so the optimum dimensions are $R = 0·71$ and $L = 1·0$. This is illustrated in Figure 5.1.

An alternative approach to pressure vessel optimization has been outlined by Cheers and Furman (1971). They suggest three possible independent criteria for the cost optimization of the vessel, namely

(a) minimum surface area,

$$D = 1·13 \, (C)^{1/3}$$

(b) minimum weld length,

$$D = \left(\frac{8.C.NC}{\Pi^2(NP+1)} \right)^{1/3}$$

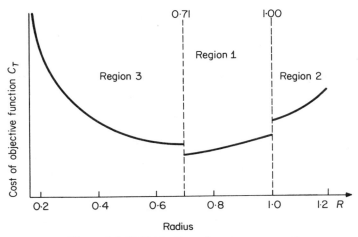

Figure 5.1 Optimization of a pressure vessel

(c) maximum plate-bending rolls capacity,

$$D = \left(\frac{4C}{\Pi.LR.NP}\right)^{1/2}$$

For each case the optimum vessel diameter D (i.e. minimum overall cost diameter) is expressed as a function of the significant variables, where C is the vessel capacity (volume), NC the number of circumferential plates, NP the number of longitudinal plates, and LR the length of plate-bending rolls. The first case applies for a design where material cost is considered to be the most significant, but this does not necessarily give the lowest tonnage nor the lowest cost of material. The second case is of importance since the welding costs usually account for over half of the total manufacturing costs. Vessel dimension selection on this basis generally leads to an improved design. The third case is of interest when, after an optimum vessel diameter has been derived from considerations of either minimum surface area or minimum weld length, it is found that the width of the plate cannot be accommodated in the bending rolls. Thus the maximum roll capacity may be the determining factor.

For correct sizing a breakdown analysis of the overall manufacturing costs of a pressure vessel is required, deriving an empirical cost equation for each item of the necessary equipment and fabrication costs involved. A search is made for an optimum. Even then an exact value is difficult to achieve because the geometry of the vessel will be influenced by availability of ends and existing designs. A non-optimum vessel in terms of fabrication costs may well prove cheaper because that manufacturer is incorporating lower overhead costs.

Similar considerations apply for most items of equipment. The objective function must be defined and the solution relates solely to a given set of input state variables. Several such sets may be fed to the optimizing algorithm pertinent to the type of unit. Eventual selection is made on an overall process basis. Note that the selection is usually not the optimum but the best attainable being limited by time, design costs, availability, and the economics of the vendor.

5.2 Logical search

The mathematical model describing the problem should be the simplest for the particular task. Thus when optimizing a process or complex it is essential to consider the role of the item in the overall economic balance. If it is crucial, considerable engineering must go into the model; otherwise use estimates or average values wherever these are adequate for the problem. Once the process is optimized a fresh look at the particular unit is necessary. This may be the designer of the process or a specialist manufacturer. Thus, in the pressure vessel example, a simple correlation may be possible in the initial model, followed by a detailed design later. The justification for the above approach is that most process calculations may be extremely repetitive and expensive. An aim is to achieve a good base case from which to consider marginal changes. Computing charges should then be smaller.

Costs may also be reduced by selecting the optimization algorithm best suited for the task. Space does not permit a discussion of the alternatives which are covered in the literature. Furthermore, many process engineers will tend to use such procedures as control algorithms provided with a process simulation package. It is considered that the following examples suffice to indicate the importance of the optimization mode. For further information a useful review has been given by Hughes (1969).

The previous examples indicate that the classical method of locating the extreme of a function using calculus is of limited value in many design problems. Fortunately some very efficient numerical search methods exist which can be used to obtain the optimum in problems which involve one or perhaps a few design variables.

The general principle of direct search is to make a number of measurements of $y(x)$, the objective function, at values of x within a region $a \leqslant x \leqslant b$. At the end of the set of measurements, hopefully the extreme value of the function will be in a certain sub-interval of the region $b - a$. Such numerical approach introduces the concept of region elimination which may be used to isolate the optimum of a function that is unimodal.

5.2.1 SEARCH BY THE GOLDEN SECTION (WILDE, 1964)

With the condition that the function is unimodal, this method involves placing two search points at a distance l from each side of region of interest of length L_K.

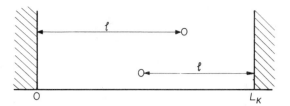

A section of the region may be eliminated by substituting these search point values into the objective function and comparing the two results. It is necessary to place these first two points so that the point remaining after the first region elimination is a fraction l of the distance from the side of the remaining region. Assuming $L_K = 1$ an approximate value of l can be derived.

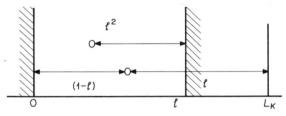

Therefore $l^2 = 1-l$ and hence $l = 0.618$. A strict derivation of the above requires a study of the Fibonaccian series of numbers.

Example 5.3. Let us now illustrate the concept of region elimination using the golden section search technique. Consider a region within the boundaries a and b, and having length L_K. We wish to place two search points $x_{1,1}$ and $x_{1,2}$ within the region so that $l = 0.618L_K$, where $L_K = b-a$.

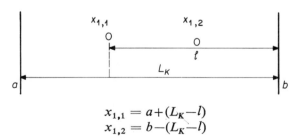

$$x_{1,1} = a+(L_K-l)$$
$$x_{1,2} = b-(L_K-l)$$

The system is resolved as follows. Assume we require a maximum (i.e. the value of x that yields the maximum value of the unimodal objective function, $y(x)$).

Cycle 1. Let $a = 0$, $b = 1$, $L_K = 1$, $l = 0.618$

$$x_{1,1} = 0+(1-0.618) = 0.382$$
$$x_{1,2} = 1-(1-0.618) = 0.618$$

If $y_{1,1}$ represents the objective function value for the point $x_{1,1}$, etc., and $y_{1,1} > y_{1,2}$, eliminate the region to the right of $x_{1,2}$.

Cycle 2. $a = 0$, $b = 0.618$, $L_K = 0.618$, $l = 0.618^2$

$$x_{2,1} = 0 + (0.618 - 0.618^2) = 0.236$$
$$x_{2,2} = 0.618 - (0.618 - 0.618^2) = 0.382$$

$x_{2,2}$ must be the same as $x_{1,1}$ since this is the point left from cycle 1 after the first region elimination. If $y_{2,2} > y_{2,1}$, eliminate the region to the left of $x_{2,1}$.

Cycle 3. $a = 0.236$, $b = 0.618$, $L_K = 0.382$, $l = 0.618(0.382)$

$$x_{3,1} = 0.236 + (0.382 - 0.236) = 0.382$$
$$x_{3,2} = 0.618 - (0.382 - 0.236) = 0.472$$

As before, one value corresponds to a point from the previous cycle, i.e. $x_{3,1} = x_{2,2}$. If $y_{3,1} > y_{3,2}$, eliminate the region to the right of $x_{3,2}$, and so forth. The calculation continues until the region remaining is within some acceptable degree of accuracy. The optimum value of x is the final region boundary value giving the larger value of $y(x)$.

The same approach applies if a minimum function value is required, the only difference being the choice of region to be eliminated after each cycle.

As can be seen, this method of search rapidly reduces the potential range of values for the independent variable under examination, and an acceptable value is quickly isolated, the only limitation to this approach of search being that the function must be unimodal.

Example 5.4. Search by the golden section for Underwood's θ. In the Fenske–Underwood short-cut method of distillation column design it is necessary to evaluate the Underwood parameter from the equation.

$$\sum \frac{\alpha_i x_{iF}}{\alpha_i - \theta} = 1 - q$$

where α_i is the relative volatility of ith component, x_{iF} the mole fraction of ith component in feed, q the state of feed, and θ the Underwood parameter. Rearranging the above gives

$$\sum \frac{\alpha_i x_{iF}}{\alpha_i - \theta} - (1 - q) = 0$$

and ignoring the sign of the function, search for the value of θ that yields the minimum value of the function.

Consider a two-component system:

$$\left. \begin{array}{l} \alpha_1 = 5.00 \\ \alpha_2 = 1.00 \end{array} \right\} \quad \text{i.e. } 1 \leqslant \theta \leqslant 5$$

$$x_{1F} = 0\cdot6$$
$$x_{2F} = 0\cdot4$$
$$q = 0\cdot5$$

Using a computer routine the values of Table 5.2 were quickly obtained.

Table 5.2. Underwood's θ by the golden section

Cycle K	L_K	Values of		Function value		Region containing
		θ_1	θ_2	$f(\theta)_1$	$f(\theta)_2$	
1	4·000	2·528	3·472	0·452	1·301	$1 \leqslant \theta \leqslant 5$
2	2·472	1·944	2·528	0·058	0·452	$1 \leqslant \theta \leqslant 3\cdot472$
3	1·528	1·584	1·944	0·307	0·058	$1 \leqslant \theta \leqslant 2\cdot528$
4	0·944	1·944	2·167	0·058	0·216	$1\cdot584 \leqslant \theta \leqslant 2\cdot528$
5	0·583	1·807	1·944	0·056	0·058	$1\cdot584 \leqslant \theta \leqslant 2\cdot167$
6	0·360	1·722	1·807	0·139	0·056	$1\cdot584 \leqslant \theta \leqslant 1\cdot944$
7	0·222	1·807	1·861	0·056	0·009	$1\cdot722 \leqslant \theta \leqslant 1\cdot944$
8	0·137	1·861	1·892	0·009	0·017	$1\cdot807 \leqslant \theta \leqslant 1\cdot944$
9	0·085	1·839	1·861	0·028	0·009	$1\cdot807 \leqslant \theta \leqslant 1\cdot892$
10	0·053	1·860	1·872	0·009	0·001	$1\cdot839 \leqslant \theta \leqslant 1\cdot892$
11	0·032	1·872	1·880	0·001	0·007	$1\cdot860 \leqslant \theta \leqslant 1\cdot892$
12	0·020	1·868	1·872			$1\cdot860 \leqslant \theta \leqslant 1\cdot880$

As can be seen in only 12 calculations the value of θ has been isolated to the region $1\cdot86 \leqslant \theta \leqslant 1\cdot88$. This compares favourably with a heuristic technique where the value of θ is initially 1 and subsequent calculations use new θ values incremented by some small interval, say 0·01, which would involve 86 evaluations to attain the same degree of accuracy.

5.2.2 SIMPLE LOGICAL SEARCH

Earlier mention is made of avoiding unnecessary detail in the model. Thus, for a heat exchanger train where individual units are unlikely to be economically crucial to the project, it is often simpler to use average heat coefficients estimated from good practice and a cost estimate correlated against function area. This reduces the computation required. Similarly an optimization algorithm such as the Hooke and Jeeves (1969) search can reduce the number of iterations through the problem.

In the following example the intention is to show the importance of the economic criteria in achieving an optimum. It is appreciated that other considerations of operating within a larger system would affect the optimum selection.

Example 5.5. Optimization of a heat exchanger stream. A heat exchanger sequence is to be designed comprising a boiler, 'stage 1', a boiler feed water

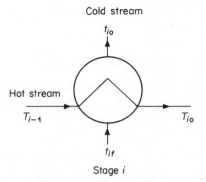

Figure 5.2 Nomenclature for stage, Example 5.5

preheater, 'stage 2', and a cooler, 'stage 3'. For simplicity let us assume that a gas stream is to be cooled from 800 to 35 °C, a flow F of 10 000 kg/h and 8400 h/annum. Defining each stage as i, heat transfer equations and data used are indicated in Figure 5.2 and Table 5.3. Design equations used are as follows. The heat transfer at stage i is given by

(1) $Q_i = U_i A_i (\Delta T_i)_{\ln}$, overall heat balance

where

$$(\Delta T_i)_{\ln} = \frac{(T_{i-1}-t_{io})-(T_i-t_{if})}{\ln\left[\dfrac{T_{i-1}-t_{io}}{T_i-t_{if}}\right]}$$

(2) $Q_i = FC_{hi}(T_{i-1}-T_i)$, balance over hot stream
(3) (a) $Q_1 = \lambda W_1$, for waste heat boiler
    (b) $Q_i = W_i C_{ci}(t_{io}-t_{if})$, for stages 2 and 3

Constraints on the design variables are

(1) $T_1-T_2 \geqslant 0$
(2) $T_0-T_1 \geqslant 0$
(3) $T_1-t_{10} \geqslant 0$
(4) $T_0-T_2 \geqslant 0$
(5) $T_2-t_{2f} \geqslant 0$

where o is output and f is input.

Cost information is available as follows. The capital cost of the heat exchangers is correlated against surface area as follows.

$$\begin{aligned}
&\text{Boiler} && I_E = 480A_1 \ (\exp 0{\cdot}48)\\
&\text{B.F.W.} && I_E = 400A_2 \ (\exp 0{\cdot}48)\\
&\text{Cooler} && I_E = 436A_3 \ (\exp 0{\cdot}48)
\end{aligned}$$

where I_E is the delivered cost (£) and A is the surface area (m). Happel's formulation is used to derive installed costs, i.e.

$$I_F = (I_E + \Sigma F_i I_E)(1 + \Sigma F_d)$$

where ΣF_i are indirect cost factors, ΣF_d are direct cost factors, and ΣI_F is the installed cost (£). Only factors which change with size of exchangers need be considered for optimization purposes, for instance instrumentation may be neglected except for the trivial case of omitting one exchanger. In this example factors were assumed as

    direct: labour 0·15, piping 0·15, foundations 0·1, insulation 0·1 (not cooler)

    indirect: overheads 0·2, engineering 0·05, contingency 0·05

The cost function U for the particular depreciation, M.A.R. is equivalent to

$$U = 0·55I_F + C_V$$

The overall cost function is given by the sum of

$$U_{boiler} = 0·060A_1 (\exp 0·48) - 0·0014 W_1$$
$$U_{B.F.W.} = 0·052A_2 (\exp 0·48) - 0·0002 W_2$$
$$U_{cooler} = 0·052A_3 (\exp 0·48) + 3 \times 10^{-7} W_3$$

where U is measured in £/h.

Table 5.3. Data for Example 5.5

Symbol	Item	Stage 1	Stage 2	Stage 3
t_{if}	cooling fluid in (°C)	240	100	20
t_{io}	cooling fluid out (°C)	240	240	35
U	overall heat transfer coefficient (kcal/m² h degC)	300	300	400
C_c	specific heat of cooling fluid (kcal/kg degC)	1·0	1·0	1·0
C_h	specific heat of hot stream (kcal/kg degC)	0·60	0·57	0·55
λ	latent heat of cooling fluid (kcal/kg)	670	—	—

The unknowns are the process stream temperatures T_1 and T_2, and if these values are defined the various flow rates will be obtained. As an interactive computer facility was available the values of T_1 and T_2 were varied over the range 800–240 and 800–100 using a simple search procedure based on the Hooke and Jeeves (1969) search plan as described in a similar example by Rudd and Watson (1968).

Following local exploration of variables, slight pattern changes were made to ensure convergence was not at a sub-optimum. The search was then reduced to using smaller incremental changes in the area of the optimum.

Defining I.C. as an incremental change, P.C. as a pattern change, B as better, W as worse, the cost analysis of Table 5.4 is obtained.

Table 5.4. Search for optimum temperatures

	I.C.	I.C.	I.C.	I.C.	I.C.	I.C.	I.C.
T_1	500	500	500	500	500	400	300
T_2	450	350	250	150	50	150	150
U	−3·07	−3·93	−4·80	−5·64		−6·03	−6·37
	B	B	B	$T_2 > 100$		B	B

	I.C.	P.C.	P.C.	I.C.	I.C.	P.C.	I.C.
T_1	200	600	320	350	250	250	250
T_2	150	450	230	150	150	120	125
U		−2·63	−5·69	−6·21	−6·40	−6·49	−6·49
	$T_1 > 240$	W	W	B	B	B	W

T_1	250	280	290	270
T_2	115	115	115	115
U	−6·48	−6·59	−6·58	−6·59
	W	B	W	W

The final values working to more decimal places are $T_1 = 278$, $T_2 = 113$, but of course the data are not sufficiently accurate to define these values as the optimum.

The system was also investigated leaving out either boiler or preheater. Some changes were made to the economic factors to permit this but the optimum remained unaltered.

Part 2

PROCESS SPECIFICATION

Chapters 6–9

In this section some background information on process engineering techniques are given. The aim is to show how process engineering is concerned with real systems and the models are developed to give a specification sheet suitable for a specialist manufacturer to produce the required item. The sheets also introduce professionalism in report writing; a calculation becomes more than a scrap of paper when the conclusions are noted in a clear manner. Similarly process-charting techniques are most important and have been discussed in detail. Process synthesis is more difficult to illustrate relying, as it does, so heavily on experience. Modifying considerations is a term for activities which essentially modify aspects of a basic design. They can have a pronounced effect on the costs.

Process charts appear extremely simple when observing the finished diagram. For the less experienced it is a useful exercise to attempt drawing out the chart from the process description and comparing one's effort with the published answer.

6 Process Charts

The use of process charts relative to the authorization stage which the project has reached is indicated in Table 6.1. The charts considered include block, information flow, process flow, process outline, activity, and engineering flow.

6.1 Block diagrams

6.1.1 TECHNIQUE

The basic block diagram is used at the start of the process design for outlining the process.

Example 6.1. Block diagram for an aromatics complex. It is possible with a specification of raw materials and products to develop a complete complex. Consider Table 6.2 which contains relevant aspects of a specification for naphtha. The aim is to maximize production of aromatics with improvement in the yield of benzene, and some production of cyclohexane. A study of the specification indicates that sulphur compounds, low boiling materials, and high boiling materials be removed and some material be converted to aromatics. After cooling, the outlet stream from the reactors contains gas and aromatic compounds mixed with other liquids. These are separated by gravity separation and liquid extraction. The aromatic fraction is further refined by distillation.

Some background knowledge is required before these considerations can be translated into the block diagram of Figure 6.1. Otherwise one can only postulate that a reaction stage such as catalytic reforming is required. Once this is confirmed the construction of the block diagram is relatively simple.

Example 6.2. Block diagram for the production of aromatics from crude benzole. In this example information is available about the reactions taking place, Figure 6.2, and details of the product specifications available (see

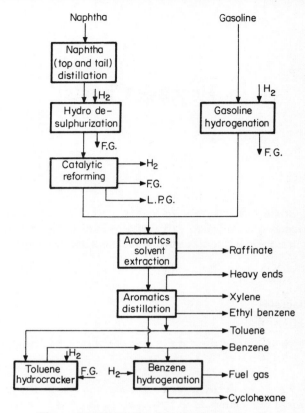

Figure 6.1 Block diagram of aromatics unit

Table 6.2 for benzene specification). Early steps are taken to eliminate compounds which do not produce benzene and consume hydrogen. In some cases it is economic to process these products to other by-products. Following the reaction the main problem is how to separate the reactants, how to decide on the products, and how to purify the gas.

For further examples see wall charts; of these Woodall–Duckam produce a superb refinery/petrochemicals example. For further information on catalytic reforming, see the British Petroleum films available from the Petroleum Films Bureau.

6.1.2 DEVELOPMENT

The process advances to the process flow diagram stage by replacing each block with items of equipment, i.e. a distillation unit becomes a column with ancillaries. This development requires a knowledge of the operations possible to separate chemicals, the properties of the chemicals which permit

Table 6.1. Process charts—use in relation to design stage

Stage	Diagram	Use in process design
preliminary	block	(i) segregate complex into plants (ii) optimize chemicals and utility requirements (iii) indicate sequence
	information/process	(i) mass and heat balances (ii) refine block diagram (iii) check feasibility (iv) check list for estimates
budget	block/information	(i) link projects (ii) preliminary site layout (iii) information flow
	process	(i) refine preliminary process charts (ii) preliminary layout
control	process/information	(i) optimize process (ii) layout
	engineering	(i) nozzles (ii) preliminary pipework (iii) preliminary checks on start-up, etc.
production	process	(i) communication (ii) control settings (iii) optimization and plant standards
	engineering	(i) model isometrics (strictly at control) (ii) check feasibility of operations (iii) check accuracy of installation (iv) detailed operation

such separation (Henlay and Staffin, 1963), and the relative cost of such operations. Although computer techniques have been suggested to reduce the labour of sifting such information (Gregory, 1968) for many purposes a very rough cost classification of common operations suffices, Table 6.3. Thus, if the operation can carry out a specific duty, a gravity separation in a tank usually is cheaper than distillation. This in turn is often less expensive than solvent extraction which could require two columns.

Heat optimization is a later feature of the design but should be considered when selecting operations. Phase change has a considerable effect on the ease of separation; gas–liquid phases are usually easier to separate than liquid–solid and movement of solids is expensive.

Unfortunately at this vital synthesis stage there is no substitute for experience and process knowledge. Computerized solutions are indeed extremely long term (Siirola, 1971).

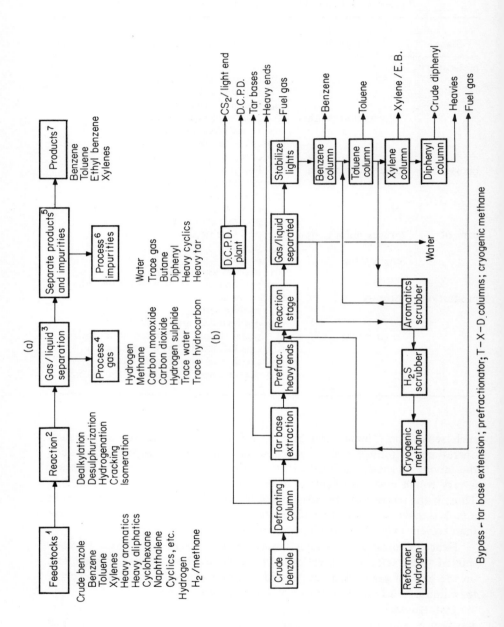

Figure 6.2 Block diagrams: (a) block diagram of reaction; (b) block diagram of process—produced by method study

Table 6.2. Material specification

(A) *Specification of naphtha*

Typically	C7 H15
Specific gravity	0·72
Boiling range	70 to 160 °C
Aromatics	20% wt.
Olefine	1% wt.
Naphthenes	25% max.
Sulphur	500 ppm wt.
Calorific value	40 000 kJ/kg
Specific heat	2·3 kJ/kg °K
Molecular weight	99
Other material is mainly paraffinic	

(B) *Specification of benzene*

Benzene content	99·9% wt.
Distillation range	0·8 °C to include 80·1 °C
Freezing point	5·45 min.
Naphthenes	0·03 max.
Toluene	0·05 max.
Aliphatics	0·05 max.
Non-volatiles	0·001 max.
Specific gravity	0·884
Sulphur	0·2 ppm wt. max.
Bromine index	20 max.

Table 6.3. Common chemical operations

(1) Gravity	(3) Electrostatic precipitation
Distillation	Ion exchange
Adsorption	Leaching
Filtration	Adsorption
Centrifuging	
Chemical conversion	
(2) Liquid extraction	(4) Other means
Crystallization	
Cyclone	
Evaporation	Use class 1 before class 2, etc.

6.1.3 MASS BALANCES

It is at the block diagram stage that a start is made on the process mass balances. As a toluene hydrocracker features prominently in succeeding chapters this is used to illustrate how short-cut procedures can be used to rough out the design.

Example 6.3. Toluene hydrocracker mass balance. In this process toluene is converted to benzene by reaction with hydrogen. Only 80% of the material is converted per pass. Methane is produced by the reaction and is also a component of the hydrogen which is obtained from a stream–naphtha reformer. To remove methane use a fuel gas purge. An overall balance may be carried out using additional data as in Figure 6.3.

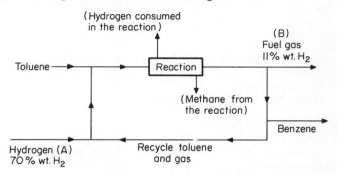

Figure 6.3 Mass balance block for toluene hydrocracker

Overall mass balance on liquid products. To calculate the recycle toluene, carry out a mass balance of toluene round the reactors.

Let make-up toluene	$= S_1$
Let feed to the reactors	$= S_2$
Then recycle toluene in reactor outlet	$= 0.2S_2$
Also feed to the reactors S_2	$= 0.2S_2 + S_1$
Therefore S_2	$= S_1/0.8$
Therefore recycle toluene	$= S_1 0.2/0.8$

i.e. the recycle toluene represents 25% of toluene feed to the process.

Overall mass balance on gaseous products. To calculate the amount of make-up hydrogen and fuel gas:

let hydrogen feed $= A$ and fuel gas $= B$
methane produced in reaction $= 100 \times 16/92 = 17.4$
hydrogen consumed in reaction $= 100 \times 2/92 = 2.2$
difference, gain in gas $= 15.2$
hydrogen balance $0.70A = 2.2 + 0.11B$
overall balance $A = B - 15.2$

and the hydrogen feed can be found by solving the two equations. This is a very preliminary statement of the mass balance. Aspects such as minor side reactions, the presence of benzene in the recycle toluene, toluene in the product benzene, and impurities in the feed must be accommodated. Furthermore, no process of this nature can exist without process losses. If the mass balance is required per hour, then the number of hours worked based on the reliability and maintenance schedules of the process is required.

Example 6.4. Distillation train design. This technique uses a short-cut procedure for preliminary screening of the idea. Consider five compounds *A*, *B*, *C*, *G*, *E* requiring four columns, I, II, III, and IV to effect separation. Two alternatives suggested are shown in Figure 6.4. Which should be

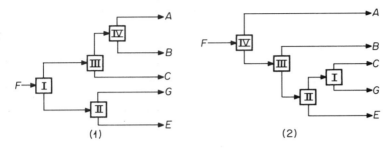

Figure 6.4 Alternative distillation trains

selected? Products *A*, *B*, and *E* are sold whilst *C* and *G* are used on an adjacent plant. The mass balance for scheme (1) is given in Table 6.4. Scheme (2) is similar. Using the well-known Fenske–Underwood system (Henlay and Staffin, 1963) calculations may be made to calculate the number of plates and reflux ratio, Chapter 7. Table 6.5 summarizes the information and thence can be calculated the overhead vapour rate. This figure gives an estimate of both the heat input and the relative capital cost of the columns, as the column heights are the same for all the schemes. There is no necessity to produce figures in specific units.

On heat input select scheme (2) assuming no problems in operability. A further advantage is column I may be installed at an adjacent plant which uses the products; this is not possible for scheme (1).

Examination of scheme (2) for heat optimization is carried out by first indicating temperature besides the streams and the heat requirement inside the column as shown in Figure 6.5.

·Table 6.4. Mass balance—scheme (1)

Column stream	I F	I D	I W	II F	II D	II W	III F	III D	III W	IV F	IV D	IV W
A	30	30					30	30		30	30	
B	30	30					30	30		30		30
C	8	8					8		8			
G	25		25	25	25							
E	7		7		7	7						

F, *D*, and *W*, refer to feed, overhead, and bottoms.

Table 6.5. Rough heat input to distillation trains

Column	N	R	Scheme (1) D	Heat	R	Scheme (2) D	Heat
I	38	7·5	68	572	64	8	511
II	46	2·3	25	83	2·2	32	108
III	335	1·2	60	135	3·0	30	119
IV	28	2·5	30	104	2·8	30	114
Total				894			852

Heat input $= (R+1)D$.

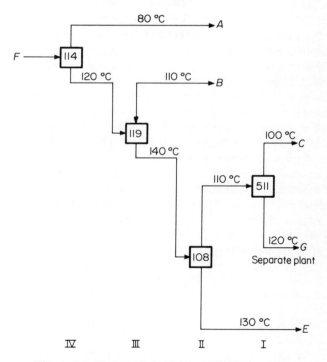

Figure 6.5 Heat optimization on distillation train

Column I is to be located at an auxiliary plant and heat in this unit can be utilized elsewhere. No column should be reboiled using the overhead from the immediately adjacent column because disturbances tend to oscillate (an exception to this would be in cryogenic distillation where such luxury of choice is not possible). However, other alternatives between the columns are possible. The overhead from column II could reboil column IV. For this to be feasible requires the temperature of the overhead stream to be raised by increasing the pressure at which this column is operated. The units nearly balance in heat requirement but excess should be provided in column II to the extent of 10% of the load in IV. Excess heat can be absorbed by the use of a trim condenser.

Such a technique is valid at stage 1 and according to circumstances may be acceptable at stage 2. The procedure can be refined by using improved techniques for sizing the columns and by calculating exact heat inputs.

6.2 Information flow diagrams

6.2.1 DIRECTED GRAPHS

The flow diagrams used to represent the relationships between units are directed graphs. Thus they consist of a collection of nodes linked by edges. If more than one edge exists between nodes i and j, a weighting w_{ij} may be given. On the block diagram the units represent nodes and the edges the process parameters. A directed graph is acyclic if it contains no cycles, a cycle being any path which returns to its starting node. Such a system represents a rapid computational scheme. An acyclic system is formed when units of the plant are arranged into a sequence of units, or groups of units, such that there is only a forward feed of information from group to group. However, it should be noted that individual units may contain many self-loops and further analysis may be required to break down these recycle streams.

For design purposes it should be noted that the flow of information is not necessarily in the same direction as flow of material, i.e. the specification of a product stream means a flow of information to a process unit as opposed to the physical flow of the stream out of the unit. An information flow diagram of a distillation system is given in Figure 6.6. The unit incorporates a column, reboiler, overhead/exchanger (E_1), overhead condenser (E_4), feed/bottoms exchanger (E_2), final feed preheater (E_3), bottoms cooler (E_W), overhead cooler (E_D), overhead pump (O), and bottoms pump (W). Instead of drawing the conventional process flow the diagram shows the path of the main information variables between the units. The subscripts are not detailed as the diagram is intended to show the type of flow rather than the process and the variable names, i.e. F feed, are referred to in Chapter 8.

Normally it is unnecessary to construct the diagram in such detail. Note the numbers inside each box per item of equipment. Mathematically a unit is composed of a set of (N) independent equations having (M) variables. Analysis involves selecting the ($M-N$) decision variables which must be fixed by the designer to leave a ($N \times N$) matrix which may be solved either simultaneously or using a feasible method for the unit. Thus, for exchanger

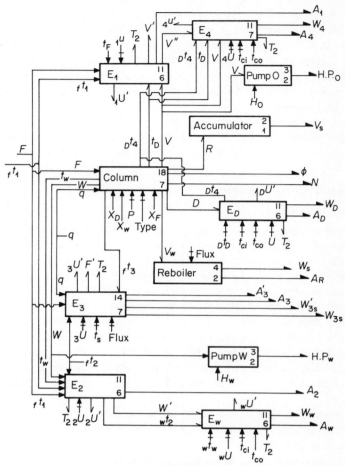

Figure 6.6 Information flow diagram, distillation unit

(E_D), $M = 11$ and $(M-N) = 6$. There are six arrows conveying information into the unit and eleven arrows associated with the box. The decision variables are assigned by the environment, e.g. mains steam pressure, cooling water temperature; others are invariant by the statement of the problem, e.g. selling specification for product; and those remaining are consumed in maximizing the profitability of the process.

Rudd and Watson (1968) assign the direction of calculation on information flow diagrams as follows:

externally specified variables	$\dashv\!\!\longrightarrow$	crossed arrows
dependent computed variables	$\longrightarrow$	half arrows
variables free to be adjusted by the designer	$\longrightarrow$	full arrows

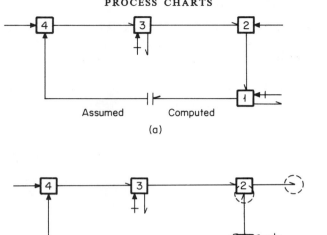

Figure 6.7 Decomposition of recycle structure: (a) tearing and iteration; (b) reversal of information flow

The individual systems are linked together consuming free variables. The resulting structure is examined. Figure 6.7(a) shows one method of decomposing the cycle. 'Tearing' involves guessing the value of an edge so as to cut an existing loop. Calculations are carried out until the values match. Alternatively reversal of flow may be feasible. In Figure 6.7(b) flow has been reversed which means a change of direction of dependent computed variables round the system until a suitable free variable at which to reverse flow is obtained. Note how there is no change in the number of arrows providing information into each unit. These represent the $(M-N)$ variables which are fixed. Note that in Figure 6.7(b) the technique employed in optimizing the system is to set the parameter at various values and check the effect on the objective function. The system is acyclic. In the system of Figure 6.7(a) the objective is agreement between assumed and computed values. In this case improved iterative procedures are desirable as successive substitution can take excessive computing time.

6.2.2 ITERATION

Consider a $f(x) = 0$.

The roots of this equation may be evolved by selecting an interval along the axis of x and evaluating at incremental values until the roots are found. Other methods (secant, Newton's) speed the convergence on the values by measuring the difference in function value or utilizing the derivative of the function to evaluate the next assumed x. If the function is placed in the form

$$x = F(x)$$

an iteration scheme can be used. Thus, after a first trial x_1, successive trials are obtained by substitution into $F(x)$, i.e. $x_2 = F(x_1)$, followed by $x_3 = F(x_2)$ . . . until agreement is reached.

Example 6.5. Iteration of a simple function. Consider the function $x^3 - 5x + 3 = 0$. Convert to the function $x = (x^3 + 3)/5$. A graph Figure 6.8 of

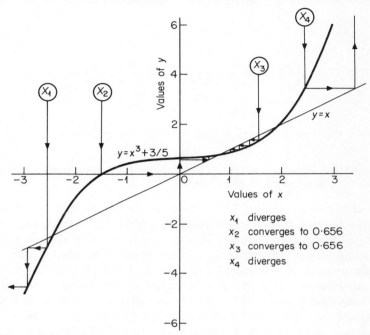

Figure 6.8 Iteration and convergence

$y = x$ and $y = (x^3 + 3)/5$ shows the roots. If a guess of x equals 0.5 were made, repeated substitution gives 0.625, 0.648, 0.654, 0.656, and the function converges on 0.656. It does so for all guesses between -2.5 and 1.8. Unfortunately for the other roots, $x = +1.9$, -2.5, it does not and diverges. The graph shows why. If the slope $F'x$ is $> |1|$ in the neighbourhood of the root the solution diverges. Preferably the mapping should be placed in a form such that for all values of x the selections converge. A suitable form of the above function is $x = (2x^3 - 3)/(3x^2 - 5)$.

Example 6.6. Convergence properties of a recycle system. In the material balance diagram Figure 6.3, toluene is recycled around the reaction system. If S_1 is the make-up stream of toluene, S_2 is the toluene in the feed to the reactor, S_3 is the toluene in the reactor outlet, and S_4 is the toluene recycle, then simplified expressions describing the process are

(1) $S_3 = 0.2 \times S_2$ (80% conversion)
(2) $S_2 = S_1 + S_4$ (no loss in distillation section)
(3) $S_3 = S_4$ (all recycled)

Try two strategies for evaluating the equations, Figure 6.9. System 1 converges but system 2 diverges. The overall equations may readily be evaluated for the alternative sequences.

System 1: $S_2 = 100 + 0.2 S_2$ System 2: $S_2 = 5 S_2 - 500$
$F'x = 0.2 (<1)$ $F'x = 5 (>1)$

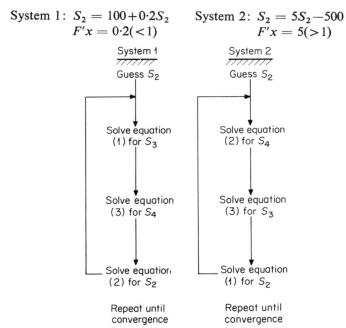

Figure 6.9 System convergence for recycle calculations

Repeated substitution of the value of S_2 in the equation

$$S_2 = 100 + 0.2 S_2$$

gives the values 100, 120, 124, 124·8, 124·96, 124·992, 124·9984, 124·999 68, 124·999 936; the system having converged when the absolute value of the fractional change,

$$\text{fractional change} = \frac{\text{new value} - \text{old value}}{\text{new value}}$$

for each element of each stream is less than a specified tolerance. The tolerance taken for the above case was 0·000 01.

If a convergence promoter is used in the system, it is convenient to describe the converged value of x by

$$x_{\text{conv}} = x_n + k_s (x_n - x_{n-1})$$

where k_s is an enhancement factor given by

$$k_s = \frac{x_n - x_{n-1}}{2x_{n-1} - x_n - x_{n-2}}$$

where n is the loop number. For the above linear system without discontinuities this promoter may be applied as a single correction after two successive iterations. Thus $x_{n-2} = 100$, $x_{n-1} = 120$, $x_n = 124$, therefore

$$k_s = \frac{124 - 120}{2(120) - 124 - 100} = 0.25$$

and

$$x_{conv} = 124 + 0.25\,(124 - 120) = 125$$

Johnson (1971) discusses the expression in his G.E.M.C.S. system and gives a computer listing. Cavett (1963) has indicated numerical methods for the convergence of simulated processes involving recycle loops.

It should be noted that iterative procedures are not the only method of generalized solution of the mass balancing problem and this section is written in the context of information flow diagrams. The generalized solution involving the simultaneous solution of linear equations proposed by Nagiev (1964) is worth further study.

6.2.3 THE ORDERING OF PROCESS CALCULATIONS

For computer-aided design an information flow diagram is produced from the process flow diagram and each line and unit (module) numbered. Note that many workers in this field have been involved in simulation rather than design and care should be taken in deciding the appropriate direction of, say, a product flow. Conventionally this is normally in the same direction as the material flow. The process, stream connection and incidence matrix are developed along the lines indicated in Figure 6.11 using a procedure appropriate to the particular system being used. This then replaces the information flow diagram. In determining equipment modules there is a one-to-one correspondence between the unit computations and the process stages. Where mixing of streams occur this may conveniently be incorporated as a junction module. Mathematical operation modules control looping or promote convergence and decision-making modules permit search for an improved solution (Crowe et al., 1969; Siddall, 1970). The addition of a costing mode is essential for design purposes. To solve a problem by computer a procedure must be determined which is unambiguous with each possible point of decision considered and the steps defined for all possible occurrences. An answer (even to the effect of no answer being possible) must be given in a finite number of calculations. Such a procedure is termed an algorithm and a block diagram of an algorithm is called a logical flow chart. To introduce use of algorithms in the ordering of process calculations see the following papers.

Sargent and Westerberg (1964) and Lee and Rudd (1966) and Christensen (1970) give algorithms for automatic generation of precedence order of calculation and locating which streams within a group one ought to tear. Johnson (1971) gives the PACER program ORDER. Thus, as shown in Figure 6.10, a block diagram may be split into groups followed by subsequent analysis.

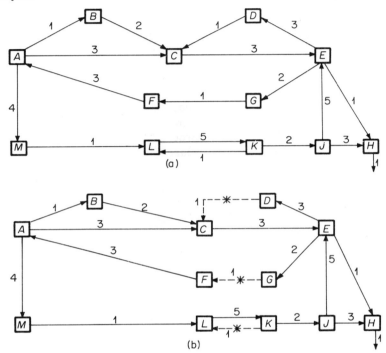

Figure 6.10 Modifying structure to organize recycle calculations: (a) initial system (containing three recycle loops); (b) structure indicating recommended tears. *Note.* Numbers indicate connecting parameters

Himmelbau (1968) has expanded the work of Harary (1959) in utilizing the drawing of a directed graph as a Boolean matrix, Figure 6.11. The subsequent analysis indicates the recycle sets.

These workers recognized that partitioning and ordering may produce computation schemes which are inefficient and unstable but at the same time their approach provides useful insight into the structure of the process and frequently achieve success.

The analysis of each unit may be expressed as one of solving sets of N algebraic equations with $M (> N)$ variables. An occurrence matrix is created which comprises columns of variables and rows of each equation, the numbers in the matrix corresponding to variable and equation. Early

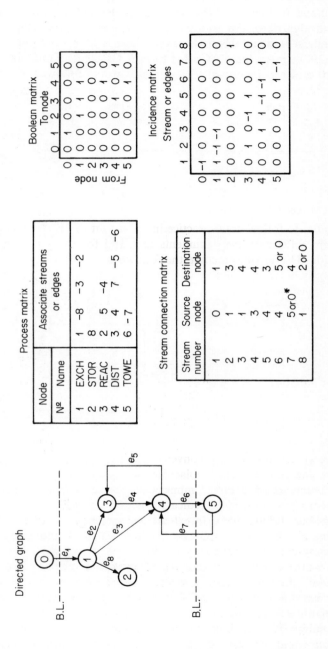

Figure 6.11 Development of matrix from directed graph. B.L. is battery limits. *O sometimes used if outside. —ve value is input to node

work by Steward (1965) resulted in locating the minimum number of tear assignments but based on structural factors rather than convergence characteristics. Westerberg and Edie (1971) have produced a criterion for output set assignment using a dynamic programming approach. The output set assignment is determined based on selecting the $(M-N)$ decision variables such that the $(N \times N)$ matrix may be solved simultaneously. If such an output set assignment avoids large eigenvalues, then convergence is enhanced. If this procedure is applied over the overall process the possibility arises of storing each unit in algebraic form, for later combination into a matrix describing the entire system. Simultaneous solution would then be attempted.

6.3 Process flow diagrams (P.F.D.)

6.3.1 RULES FOR CONSTRUCTION

It is essential to follow good practice in the layout of process flow diagrams (see Figure 6.12) as these documents are vital for the communication of information. Useful rules include the following.

(a) Symbols for equipment to conform to the British Standard (British Standards Institution, 1953) or to be in sufficient general use to be easily recognized.

(b) Raw materials enter on the left and products leave the process on the right of the drawing.

(c) Equipment to roughly scaled in proportion and at approximately the right elevation. For convenience some slight departure from this latter rule is permissible.

(d) Preferably the drawing should be on the same piece of paper.

(e) Pressures, temperature, flow rates, and heat inputs to be indicated adjacent to the lines, inside the appropriate symbol as shown by the key.

(f) Mass balance information may conveniently be listed along the top of the drawing. This is more convenient for checking and later operation; along the side permits easier replacement and upgrading of information.

(g) Streams to be identified numerically inside a box on a line.

(h) Instrumentation to be indicated using conventional symbols (British Standards Institution, 1964). In extent this relates to the time in the life of the project at which the P.F.D. is produced. It is recommended that only the main controllers and their valves should be shown. No number or indication of whether the instrument is in the control room is necessary. Alternatively the P.F.D. can be updated as information becomes available. By this means in the design part of the examination of the Institution of Chemical Engineers more information is included on the P.F.D.

(i) Equipment should be numbered using symbols appropriate to the type of item and the series number of the plant. Common labels include: B, fired heaters; C, compressors; D, convertors and towers; E, exchangers; F, tanks; G, pumps. These labels also appear on the sheet for evaluating

Stream number		1		2		3		4		5		6		7		8		9		10		11		12		13		14		15	
Stream name		Toluene feed		Recycle toluene		Vaporiser liquid		Hydrogen feed		Vaporiser gas		Quench gas		Reactor feed		Reactor prod'ts		Fuel gas F101		Recycle gas		Stab. fuel gas		Fuel gas		Stabilizer feed		Benzene tower		Benzene product	
Component	Mol wt	lb mol/h	lb/h	lb mol/h	lb/h	lb mol/h	lb/h	lb mol/h	lb/h	lb mol/h	lb/h	lb mol/h	lb/h	lb mol/h	lb/h	lb mol/h	lb/h	lb mol/h	lb/h	lb mol/h	lb/h	lb mol/h	lb/h	lb mol/h	lb/h	lb mol/h	lb/h	lb mol/h	lb/h	lb mol/h	lb/h
Benzene	78	0.4		0.4	28		28							0.4	28	244	19028						100				19028	243	18928		18900
Toluene	92	244	22400	60	5592		27992							304	27992	60.8	5600									60.8	5600	60.8	5600		8
Heavy avs	110			7.6	830		830							7.6	830	7.6	830									7.6	830	7.6	830		
Total liquid		244	22400	68	6450		28850							312	28850	312	25458										25458	311.4	25358		18908
Hydrogen	2							635	1270		2480	560		1520	3040	1276	2552		712		1770		70		782		70				
Methane	16							35	550		8640	1940		662	10580	906	14484		4054		10030		400		4454		400				
Total gas								670	1820		11120	2500		2182	13620	2182	17036		4766		11800		570		5236						
Total flow		244	22400	68	6450		28850	670	1820		11120	2500		2494	42470	2494	42494		4766		11800		570		5236				25358		18908

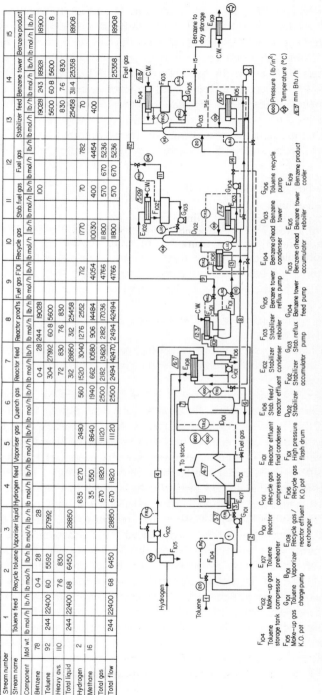

Figure 6.12 Process flow diagram, toluene hydrocracker

the capital cost of the plant. If the plant is part of a complex which is given the number 3, then the twelfth exchanger on that plant is E-312. Similarly the drawing itself should have a number and issue.

(j) It is convenient at the process flow diagram stage to show the correct stream passing through the tube side of heat exchangers. Table 6.6 indicates a general pattern, whichever fluid coming higher in the list being selected to pass through the tubes. Obviously, if later the design is modified a change must be made.

Table 6.6. Fluid through tubes in heat exchanger design

Select whichever fluid is higher in table:
    cooling water
    corrosive fluid
    one depositing solids or fouling
    fluid of lowest viscosity, if abnormally viscous fluid present
    the fluid under highest pressure
    the hotter fluid
    the liquid of less volume

Normally pass condensing vapours through the shell except for condensing steam which is passed through the tubes

---

6.3.2 PROCESS EXAMPLES

Example 6.7. Toluene hydrocracker. The process has been described earlier. Features to note are the efficient heat recovery, separation of gas and liquid by gravity in a high-pressure flash drum, and recycle of unconverted toluene and hydrogen. Note how it is unnecessary to change process heat into steam in order to use it elsewhere in the process. In this case the process stream is taken to the reboilers. A further aspect is the omission of storage between two stages of the process (reaction and distillation), the saving in capital being greater than increased costs stemming from lower reliability.

The chart is optimized by the appropriate techniques, in particular using information flow diagrams and the techniques outlined in Part 3. Although the process flow diagram would be in the operating manual, etc., it is of course reasonable that the drawing be further modified to expedite the project, e.g. show piping size and materials of construction thereby assisting personnel such as the layout engineer. The drawing may then become a process and instrument (P. and I.) diagram.

Example 6.8. Carbon dioxide removal plant. This process is described in some detail to permit the reader constructing the process flow diagram.

Potassium carbonate solution is used to remove carbon dioxide from reformer gases. The overall reaction describing the mechanism of the scrubbing process which is accomplished in an absorption tower packed with several beds of Intalox saddles is

$$K_2CO_3 + CO_2 + H_2O \rightleftharpoons 2KHCO_3$$

The reaction is exothermic as written above. Carbon dioxide is removed from 25% to 2%. The absorbed carbon dioxide is removed by heating in a stripping column and the solution recirculated. As the reformer gases contain some water this must be controlled at a specific level. The operating pressure of the absorber is considerably greater than the stripper. Circulation of carbonate is carried out by semi-lean and lean pumps. The main portion of the carbonate solution (some 75%) is leaving the stripper at a point where complete regeneration has not yet been achieved. It is fed into the absorber one-third of the height below the top. The lean solution from the bottom of the stripper is pumped to the top of the absorber. To prevent carry-over of carbonate solution a de-mister with flush connection is installed in the top of the absorber whilst any further entrained particles are removed in the carbon dioxide absorber product knock-out drum. Carbon dioxide leaves the stripper saturated with steam and so must be cooled before venting and a portion of the condensate recirculated.

The carbonate solution contains 30% weight/weight equivalent potassium carbonate plus some additive. Low temperatures favour the absorption of carbon dioxide and for this reason lean solution is cooled before it is fed to the top of the absorber. Provision is made at shut-down to store the entire solution in the absorber and stripper. Carbonate is delivered in bags and must be dissolved for make-up and initial system inventory. The P.F.D. of this process is given in Figure 6.13.

Example 6.9. Solvent crystallization. Solvent crystallization in this context is crystallization of a solution followed by the addition of liquid (solvent) to improve the eventual purity of the product by permitting better separation of crystal and mother liquor (Cutts and Wells, 1971). Routes may be synthesized by examining each essential step and generating options.

(i) To change the phase from liquid to solid.

(a) Cool using an external coil and an internal agitator.

(b) Cool using an internal coil and agitator.

(c) Use a scraped surface exchanger.

(d) Expand a boiling liquid.

(e) Cool using cold immiscible gas, liquid or solid.

(ii) Let us assume controlled crystallization then takes place.

(iii) To separate solid and liquid use the following techniques either separately or jointly.

(a) Gravity; (b) filtration; (c) centrifugation; (d) displacement by a miscible liquid; (e) displacement by a liquid product; (f) displacement

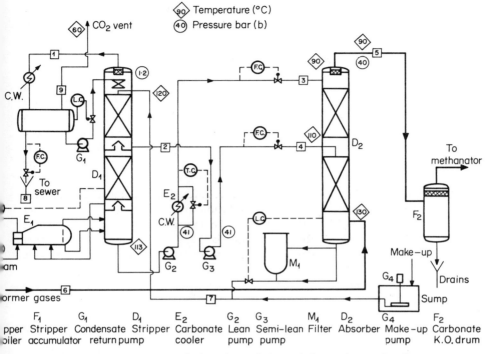

Figure 6.13 CO$_2$ removal plant (mass balance information omitted)

by an immiscible liquid or gas; (g) adsorption; (h) diffusion; (i) drying; (j) freeze-drying.

The above operations may now be linked together in a search for the optimum process for a particular separation problem.

Several well-known processes illustrate the above principles.

(i) Desalination—U.K.A.E.A. (Simon-Carves, 1970). Phase change is achieved by the expansion of liquid butane (pressurized gas) the resulting ice-brine slurry is washed with the product—water.

(ii) P-xylene—Phillips (McKay, 1969). A scraped-surface heat exchanger is used to form the crystal slurry which is washed by the products.

(iii) Naphthalene purification (National Coal Board). The feed is chilled and crystallized, followed by leaching with multiple washes of methanol–water. Final separation is by gravity and centrifuge.

(iv) Schildknecht column (Henry, 1969). The feed is chilled and crystal-lized in a scraped-surface heat exchanger, then pushed down the column by a rotating spiral countercurrent to the reflux liquid, which is produced by melting the crystals at the end opposite to the freezing section.

(v) Benzene purification (French, 1959). The feed is frozen by a cold brine stream and centrifuged with partial brine thaw. Figure 6.14 gives a

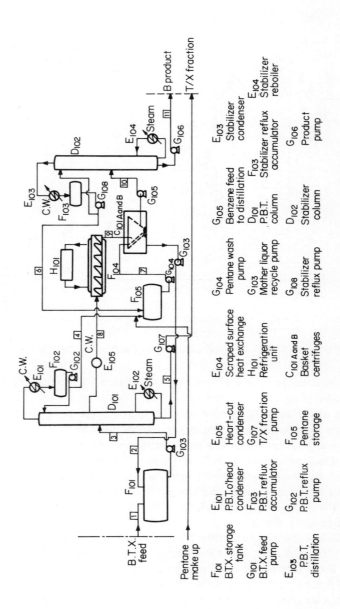

Figure 6.14 Solvent crystallization line diagram. Production of benzene from B.T.X. stream using pentane as solvent

suggested line diagram for solvent crystallization of benzene using a pentane wash. There is no doubt that this is uneconomic but it illustrates the technique of synthesizing processes. The feed is crystallized by contact with a cold gas followed by a wash with a miscible liquid.

Example 6.10. Cryogenic plants. The importance of heat recovery is shown by this example of a cryogenic process. Recovery of hydrogen from a process gas is effected by freezing out impurities such as methane. The gas is cooled until it liquefies and gravity separation removes liquid from gas. Complexity arises from the low temperatures and the desire for highly efficient heat exchange. The flow sheet Figure 6.15 indicates one of the ways Petro-carbon Limited have overcome some of these problems.

The feed gas stream free of trace impurities enters the main part of the process where it is cooled in a series of heat exchangers. As the temperature of the gas falls, condensation starts and proceeds progressively. The condensed liquid is separated at selected temperature levels and re-evaporated to cool the incoming feed gas. The limitation on this cooling procedure is the solid–liquid equilibrium characteristics. The main impurities present in the feed gas are removed from the process as a tail gas stream containing hydrocarbons and some hydrogen.

Some refrigeration is required to overcome heat losses and in certain cases adverse enthalpy differences between feed and product streams. The method illustrated is that of the cold nitrogen cycle. Full use is made of the refrigeration capacity of the condensed components by expansion to low pressure.

Note the dashed outline of the cold box which is all that the purchaser sees of the heat exchangers, vessels, pipework valves, expanders. The heat exchangers, small vessels, and pipework are constructed from high-grade aluminium alloys. The cold box consists of a structure to support the equipment, to hold the powdered insulant in position, and to maintain the necessary inert atmosphere in the insulation space.

The plant illustrated can process feed gas from a catalytic reformer. This gas may contain impurities such as water, hydrogen sulphide, ammonia, and carbon dioxide which must be removed before the gas is cooled. At these operating temperatures they will freeze in the solid phase, blocking the heat exchangers. Adsorption on beds of molecular sieves is one method of removal. If heavy hydrocarbon oils are present, these must be removed by an oil wash.

6.4 Method study techniques

For batch processes a process flow diagram gives little indication of how the process is operated. For instance, a batch distillation column in a process flow diagram gives little indication how it operates. Similar problems arise in

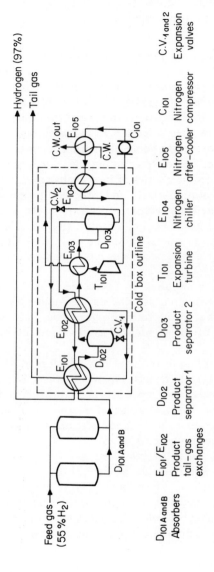

Figure 6.15 Cryogenic hydrogen recovery plant

processes. Thus, to assist in design and to provide information for operators, process outline charts may be constructed.

Five symbols are used to convey information:

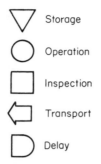

Storage

Operation

Inspection

Transport

Delay

Operations are usually numbered to assist in optimization.

Example 6.11. Sorbitol process. Construct a process outline chart for the following process. Bags of sucrose are slit and the powder conveyed into a disolving tank, 2000 gallons, in which a 50 wt. % mixture is made with distilled water. The tank is emptied into one of two similar tanks which are used to charge any of four autoclaves in 1000 gallon batches. A catalyst slurry is discharged into the autoclave from a drum and the sucrose charged. The unit is pressurized with hydrogen, pressure being maintained at 400 lb/in^2 during the reaction in which hydrogen is absorbed, thus producing sorbitol. After up to 24 hours of reaction at 150 °C, the solution is discharged into a tank. From this tank liquor is circulated through a plate and frame press using fuller's earth in a sorbitol solution. Cake is discharged to truck and the filtrate passes into one of two 3000 gallon tanks used to feed the de-ionizing columns. These remove trace impurities and are regenerated by acid, caustic, and water washes. Product is discharged into one of two 3000 gallon tanks. The 40 wt. % material is concentrated in a continuous evaporator to 70 wt. % final product and passes into a make tank for checks on product quality.

It is necessary to incorporate further tests from those indicated. An outline chart of the process is given in Figure 6.16.

To permit sizing of equipment in a predominantly batch system two approaches may be adopted.

6.4.1 ACTIVITY CHART

Activity charts such as outlined in Figure 6.17 are used to permit operation of the equipment to be studied. Activities may be grouped into operation (coloured red), delay (yellow), and inspection (blue). The aim is to improve the productivity of the equipment. Magnetic strips are extremely useful and additional symbols may be incorporated to permit the level of material in

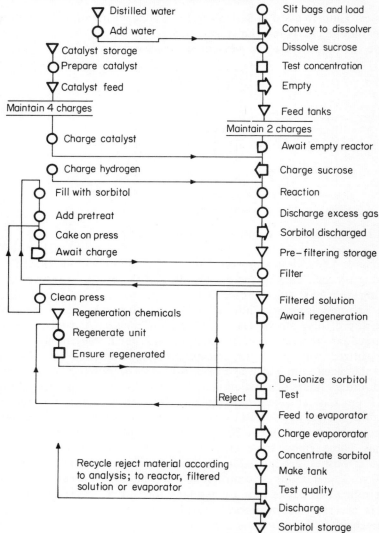

Figure 6.16 Process outline chart, sorbitol plant

the tanks to be noted. To allow for non-uniformity of, for instance, reaction time alternative sequences should be examined.

6.4.2 COMPUTER SIMULATION

An information flow chart for dynamic simulation of the process is developed. If the process enters a continuous operation such as evaporation in the above example, this is interpreted as a steady state slow whilst operating and can be

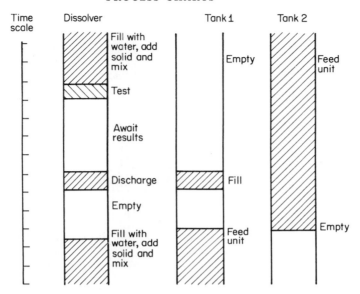

Figure 6.17 Activity chart

considered to take no time. Special computing languages are useful such as control and simulation language (C.S.L.) which contains a series of test statements, any activity only being carried out when all the tests are satisfied. In a discrete state model only those times at which an activity can occur need be considered. Time cells are specified and, as the simulation proceeds, values representing the time that must elapse before a change can occur are incorporated. Overall time is represented by a variable CLOCK. The activities are listed and the program run. If a delay or failure occurs, such as attempting to overfill a tank, an error signal is noted. The process is then modified. Variation of, say, reaction time is allowed by evolving a frequency distribution and entering data by a random sampling technique.

6.5 Engineering flow diagrams (E.F.D.)

The engineering flow diagram (E.F.D.), Figure 6.18, contains every process valve and instrument. Basically it is an extension of the P.F.D. but in such detail that several diagrams are necessary to cover the entire number of plant items. Thus numbering of diagrams is most important.

Drawing conventions vary widely. Commonly accepted procedures include the following.

(a) Indicating main process lines by heavy continuous lines.

(b) Indicating service and minor process lines by normal continuous lines.

(c) Noting on each line reference numbers giving the diameter, material of construction and identification number (for clarity omitted in Figure 6.19).

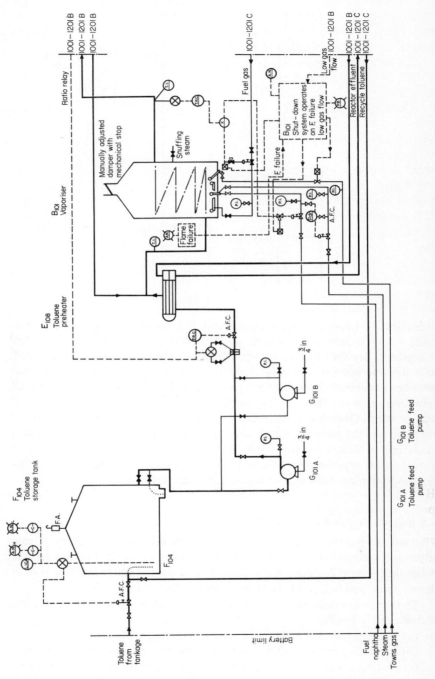

Figure 6.18 Engineering flow diagram, vaporizing section, toluene hydrocracker

(d) Showing control signal lines as dashed lines. Instruments are shown by a circle enclosing a code word, with a bar indicating whether this is displayed in the control room. Transmitters are normally shown.

(e) Indicating floor levels and if required correct vertical elevations.

(f) Labelling plant items as in the specification sheets and process flow diagram.

Certain items arise repeatedly including double block, bleed, and by-pass around a control valve to permit emergency maintenance; twin valves on vents for high-pressure lines; pump and stand-by with valves and instrumentation; instrumentation such as high low-level annunciators with associated level controllers; where controllers are fitted installing an additional measuring device, i.e. a level gauge by a level recording controller, to permit continued operation if the instrument becomes defective; relief valve locations. It is useful to prepare separate block diagrams showing the emergency shut-down systems on the plant. These are also noted on the E.F.D. but tend to lack clarity when spread over several drawings.

Each aspect of a system should be tested by a mental imposition of strains. Such actions for normal and emergency start-up, normal and emergency shut-down, control failure, etc., would appear fairly obvious, e.g. if a cold reactor feed-stream is heated up to temperature by the hot outlet stream from the same reactors, this may be feasible when equilibrium is reached but how does the unit start up? Of particular value can be imposing slight changes to control readings and deciding the action to be taken by the operator. Check all safety aspects. Armistead (1959) has classified some direct causes of destructive accidents, principally involving fire. They include the following.

(1) Misoperation or improper practices.

(2) Equipment failure.

(3) Repairing equipment when operating.

(4) Lightning, wind-storm, and other effects of the elements.

(5) Improper equipment.

Hudson (Fawcetts and Wood, 1965) has indicated check lists for different processing sections.

It can be helpful to use a specification sheet, Figure 8.1, when producing an engineering flow diagram. The list of nozzle connections provides a valuable check list. Conversely the specification sheet should be completed in conjunction with the E.F.D.

7 Aspects of Instrumentation and the Process Charts

7.1 General

It is not intended to discuss the design of instrumentation and dynamic simulation has been omitted. Instead we will discuss some aspects with which every process engineer becomes involved. These include the following.

(a) Where should instrumentation be provisionally located on the process flow diagram?

(b) How many variables should be controlled on a given system?

(c) Can the operator readily understand the effect of changes on the system?

(d) Can the operator manipulate the instrumentation if this becomes necessary?

(e) What emergency shut-down systems have been installed?

(f) Should control valves be installed with by-passes?

(g) Are sufficient relief valves installed and should there be a flare system?

7.2 Preliminary instrument location on the P.F.D.

Although sizing of instruments is well covered in the literature, less emphasis is made on locating instrumentation as on the process flow diagram. Final location is clearly a function of detailed instrumentation design, but, as a guide to the instrument engineer, the process engineer will locate equipment on the flow sheet. It is as well to do this correctly. There is, of course, considerable difference between

one controlling the back pressure on the process and the other releasing material into a stream as the pressure so demands. The second system instead of the first could lead to some rather startled operators as relief valves blast off.

Example 7.1. Control of a distillation column

(i) *Variables for distillation control.* Distillation control presents problems of selection due to the range of variables. Bauer and Orr (1954) have identified these very clearly by reference to the construction of McCabe and Thiele (1925).

(a) *Equilibrium.* For a binary system it is possible to draw an (x, y) diagram at constant pressure P. P

(b) *Mass balance.* If we know x_F, x_D, x_W, and F, D and W are automatically generated.

i.e. $F = D + W$ 6 variables $\left.\begin{matrix} \\ \\ \end{matrix}\right\}$ 4 degrees of $\qquad \left\{\begin{matrix} x_D \\ x_W \\ x_F \\ F \end{matrix}\right.$
     $Fx_F = Dx_D + Wx_W$ 2 equations $\left.\begin{matrix} \\ \\ \end{matrix}\right\}$ freedom

(c) *McCabe–Thiele construction.* On the (x,y) composition diagram shown in Figure 7.1, proceed as follows.

(1) Fix the q line. q

(2) Assume the column has a fixed number of trays and the feed $\left\{\begin{matrix} q \\ N \\ N_F \end{matrix}\right.$
tray is also fixed.

By trial and error draw the rectifying section and stripping section operating lines so that the number of trays and the feed tray fit exactly. As this is feasible the system has been fully defined by fixing the 8 variables

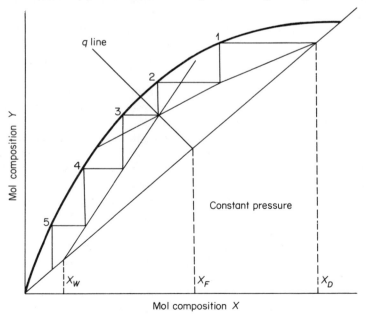

Figure 7.1 McCabe–Thiele construction for distillation design

listed in the margin. The construction assumes that reflux temperature is controlled, making nine variables. We shall neglect column internals as t_R fixed by the design.

(ii) *Selection of variables for control purposes.* The above method only indicates the number of variables which are to be fixed. It is more convenient to select the variables which are readily controlled.

(1) Feed tray—it is usual to install 3 feed locations up a tower. This permits only a small range of variations and is virtually fixed. It becomes important if 'pinches' are detected during operation.

(2) q line—adjustment is possible by appropriate heat exchange and sometimes controllers are fitted (Williams, 1961). However, in general this is fixed by the design.

(3) Pressure—only minor changes are possible once the design is complete. Take care that, if design temperatures are used as a check on composition, the design pressure is used.

(4) Number of trays and column internals—fixed by design.

(5) It is assumed that column reflux temperature is maintained at a value which does not upset the operation of the column, and that tray efficiency and other dependent variables of internal column operation will not vary greatly within the constraints of the operating range.

(6), (7), (8), (9) Four variables to be chosen from, say, F, D, W, x_D, x_W, V', L, x_F.

With regard to the original design selection of 'q' this is mainly a function of heat distribution over preheater and reboiler. Heat introduced at the reboiler is more efficient as it increases the vapour rate on all the trays, whereas the preheater increases the vapour rate in the rectifying section only. On the other hand, the cost of preheater is less than the reboiler, particularly if the feed inlet point is low.

Any distillation system must be a determinate system and one stream must be quantity controlled, i.e. flow-recording controller on steam to reboiler—hence control of the vapour flow, V'.

Composition control is necessary as owing to marginal changes in operating conditions so composition of distillate and bottoms can change, i.e. thermal conductivity, specific gravity, boiling temperature, chromatography.

Four schemes will now briefly be discussed (Williams, 1961) to give an indication of the scope, Figure 7.2. In system 1 feed composition is assumed fixed; the system has no quality control and the operators have to adjust the flow controllers should any change occur.

In other words if the assumption is wrong we can control N, N_F, F, q, D, P, t_a (within small range), V' (F.C. onstream) and the control of the last variable, x_D, becomes manual, the 'monkey on a stick' principle. Alternatively, as in system 2, a temperature composition control can be introduced on the reflux permitting feed composition to vary. Note that the composition of x_W will still adjust, and priority is given to x_D. However,

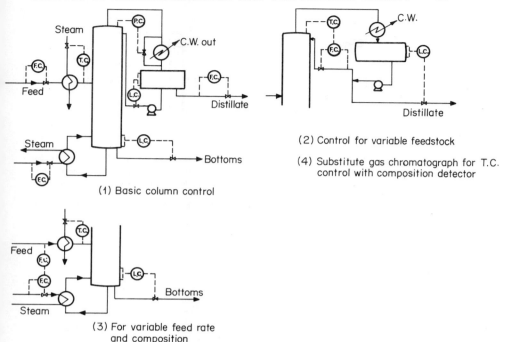

(1) Basic column control

(2) Control for variable feedstock

(4) Substitute gas chromatograph for T.C. control with composition detector

(3) For variable feed rate and composition

Figure 7.2 Some distallation column control systems

feed rate is not free to vary as is required for, say, processing the bottoms product obtained at a rate determined by level control from an earlier column. System 3 permits flexibility on this account, resetting the flow control on steam to the reboiler by a cascade controller operating from feed input rate measurements. Alternatively a temperature control operating below the feed tray could reset this steam flow. As temperature measurements do not relate to design composition unless pressure is constant and are not particularly sensitive to small change in composition, then more accurate composition control might be required as suggested for system 4.

Level controllers are only required to ensure that column or accumulator does not fill/empty. They do not control the system itself.

Controllers should be placed with the measuring element in a position such that a reasonable range of changes might be detected. A temperature controller on a column overhead detects little change in composition when producing pure materials.

(iii) *Example 7.2. Operability of a distillation system.* The process engineer must also consider the operability of the system. Consider the problems eventually facing the operator, when designing a system of columns separating a 40/20/20/20 mixture of cyclohex*ane,* cyclohexan*ol,* cyclohexan*one,* and

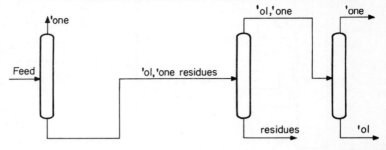

It is important that no cyclohexane is in the bottoms product.
A small amount of cyclohexanol in the recycle stream does not matter but should be restricted. Thus all operations are designed to ensure that bottoms control takes preference

Bottoms should be limited in 'ol/'one, but 'ol/'one should have no residues in them, so preference given to overhead product

Both are products so they require close control: this makes operation difficult as if the column is upset which product do you save

Alternatively

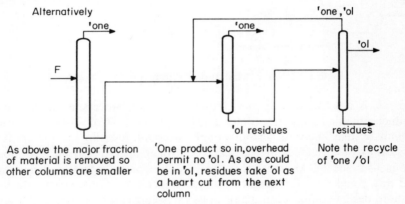

As above the major fraction of material is removed so other columns are smaller

'One product so in,overhead permit no 'ol. As one could be in 'ol, residues take 'ol as a heart cut from the next column

Note the recycle of 'one /'ol

Figure 7.3 Operability of distillation trains

residues. 'Ane is recycled back to the reaction stage and 'ol and 'one are products. The system and various considerations are identified in Figure 7.3.

What does the operator do when a stream is off specification? Does he ring up the design department and ask them to sort it out? Obviously he must think why his product is off specification from a study of the temperature profile down the column.

In Figure 7.3 consider the 'one/'ol residues, second column. Let us assume quantity control on feed and reflux, and composition control on a plate in the section below the feed tray, using a temperature-recording controller which resets steam to the reboiler. Temperature recorders are fitted at the top and bottom of the column.

(a) If the overhead temperature is high and the bottoms temperature and bottom control temperature are normal, what is happening? Too much 'ol is being pushed up the column, so reduce the setting of the boil-up temperature controller.

(b) If the overhead temperature is high and bottoms temperature low, what action is taken? Inefficient separation and contacting through the column, so increase the reflux.

(c) If the top temperature is normal and bottoms temperature low, how can the operator correct? Too much 'one in bottoms. Increase boil-up rate, but slowly, taking care not to upset the overhead stream.

(d) What if overhead temperature is low and bottoms temperature normal? One can increase the boil-up rate but check the previous column. Is some 'ane getting through? Check by analysis.

7.3 Emergency shut-down systems

On the E.F.D. of Figure 6.19 a shut-down system was indicated for a vaporizer. This system closed down the unit if there was low gas flow or flame failure, or the operator pressed the 'panic button'. Such a system is obviously required for such a unit and the process engineer must advise accordingly. Figure 7.4 indicates a more complex trip system for a hydrogen plant.

All these trip systems are eminently desirable and recommended. However, there is one snag. If the electrical system is not well designed, one can get

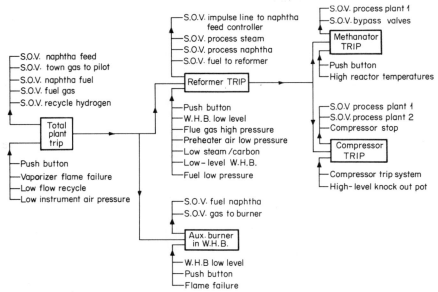

Figure 7.4 Trip system: S.O.V., shut-off valve

occurrences such as a thunderstorm causing a voltage dip and all the trips dropping out even though no other equipment failed. Similarly, carbon has been known to fall on the face of photo-electric cells with the result that the system sees a flame failure. Such occurrences can result in everyone being somewhat upset. The compromise solution is to think really hard before deciding on a trip system. A warning light, after all, does give one time to react. To reduce spurious tripping and malfunction, make provision for on-line trip testing. Usually each trip channel, the interposing relay, and all indication circuits are operated during such a test but not process valves (Frankton, 1971).

7.4 By-pass around control valves

In the studies on E.F.D. reference was made to double block, bleed, and by-pass on control valve systems. Such requirement is based on the engineering reliability of this component. If reliability R is measured by the number of past successes divided by the number of trials, then for a system consisting of three operations connected in either

Series

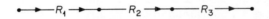

Stage 1: No. of successes $= R_1 T$
Stage 2: No. of trials$=$No. of successes stage 1 $= R_1 T$
Therefore   No. of successes at stage 2 $= R_1 R_2 T$
Stage 3: No. of successes $= R_1 R_2 R_3 T$

where T is the number of trials. As the number of successes at the last stage determines those of the entire system, system reliability $= R_1 R_2 R_3 T$.

This assumes that the reasons for failure of each component are independent of one another.

Parallel. For a parallel structure such as two pumps, one being a spare, feeding a reactor,

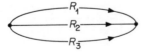

Stage 1: No. of successes $= R_1 T$

When stage 1 breaks down use stage 2.

Stage 2: No. of trials $= T - R_1 T$
No. of successes $= T(1 - R_1)R_2$
System$_{1,2}$ No. of successes $= T\{R_1 + (1 - R_1)R_2\}$
$= T\{1 - (1 - R_1)(1 - R_2)\}$

Stage 3: No. of trials $= [T-T\{1-(1-R_1)(1-R_2)\}]$
No. of successes $= [T-T\{1-(1-R_1)(1-R_2)\}]R_3$

System$_{1,2,3}$ No. of successes
$$= T\{1-(1-R_1)(1-R_2)\} + [T-T\{1-(1-R_1)(1-R_2)\}]R_3$$
$$= T\{1-(1-R_1)(1-R_2)(1-R_3)\}$$

For a system consisting of three operations which have succeeded 99/100, 98/100 and 97/100 times in the past, the chances of simultaneous success of all three is

$$\text{series:} \quad R_s = R_1 R_2 R_3 = \left(\frac{99}{100}\right)\left(\frac{98}{100}\right)\left(\frac{97}{100}\right) = \frac{96\cdot4}{100}$$
$$\text{parallel:} \quad R_p = 1-(1-R_1)(1-R_2)(1-R_3) = 1-6\times10^{-6} = 1$$

For control valves the figure will be close to 100 % success. If firms measure the reliability of the valve in a specific service, they can balance the additional cost of by-pass against the savings in on-stream time. Remember too that the block valves have a reliability as well. A hydrogen plant we were commissioning had to be shut down because of a leak in the block valve installed to permit removal of the control valve.

The present practice is to use experience in determining which control valves to safeguard in this manner. However, firms should keep records of the reliability of valves in specific duties so that correct design procedures may be used in the future.

7.5 Relief valves

The process engineer requires a knowledge of where to locate relief valves on engineering flow diagrams. To safeguard increase in pressure in the event of a fire, only if the items are not separated by valves is one relief valve required. If catalyst may crush and result in a blockage, clearly further relief is needed. Similarly, relief is needed on some liquid filled lines which can be blocked in and then exposed to the sun. Process hazards arise if the cold side of an exchanger can be filled and blocked in; relief is needed on the cold side. Blocked outlets on reciprocating compressors and pumps, gear pumps, steam exhaust side of turbines, and fractionating columns can result in raised pressure. This might be the fault of control valve freezing, i.e. reflux failure. Abnormal conditions might require additional relief. These could stem from equipment failure or water hammer. Vacuum relief is needed on tanks to safeguard when pumping more liquid out than entering or when injecting cold liquid after steaming out a vessel.

The valves normally discharge to atmosphere at a height of 10 feet above the highest working level or roof within a radius of 40 feet. Exceptions to this are when relief valves discharge the following.

(a) Heavy hydrocarbon vapours having a molecular weight of 60 or more.

(b) Toxic or lethal vapours when in concentration.

(c) Flammable liquids.

In this case relief valves discharge to a closed system leading to a flare or blow-down drum.

8 Equipment Specification and Selection

8.1 General

For the detailed design of equipment the process engineer will use texts and programs with which he is familiar. However, published sizing techniques often stop short of complete specification and for this reason it is useful to indicate typical procedures.

8.2 Data sheets

A specification data sheet is given in Figure 8.1. The data provided on the form may be complete or partial depending on agreement between supplier and client. Apart from its importance in transmitting information to a manufacturer it is a document providing reference to flow sheets and final design. The upper part is mainly process, the lower part mechanical.

8·2.1 PROCESS SPECIFICATION

The process information is obtained as the process flow diagram is developed. For the design sequence the data sheet suffices as a check list, the process engineer considering in turn aspects such as the following.

 (a) Process design and design criteria.
 (b) Type of equipment available.
 (c) Codes of practice.
 (d) Materials of construction.
 (e) Vessel geometry, approximate cost, and preliminary optimization.
 (f) Mechanical attachments.
 (g) Nozzles, main process, and subsidiary connections.
 (h) Instrument connections.
 (i) Safety devices.
 (j) Means of access into vessel, closures, etc.
 (k) Supports.
 (l) Testing, stress relieve.

Client: P.E.W.E.O.
Unit: Hydrocracker
Equip. name: Benzene column

See example for details.
Location:
Item No: D-103

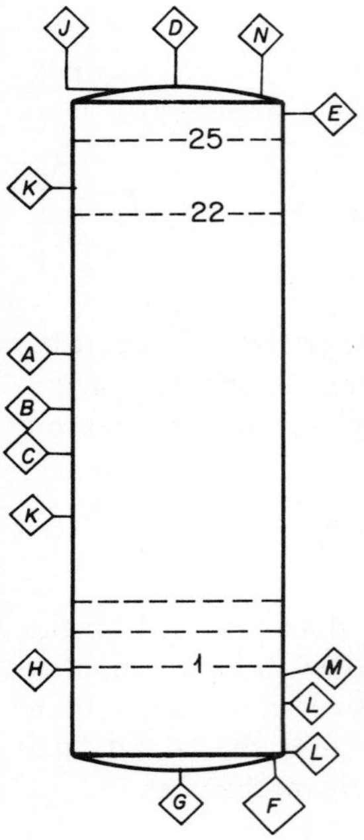

		Top	Bottom
Shell, outside diameter (m)		1·50	1·50
No. of trays		25	25
Pressure (b)	Operating	2·3	2·7
Temperature (°C)	Operating	95	130
Material	Shell	A.S.T.M. A285 'C	
	Cladding	A1	
	Trays	S.S.	
	Caps	S.S.	
Corrosion allowance (mm)	Shell	2·0	
	Heads	2·0	
	Trays	NIL	
Tray spacing (mm)		450	
Type of liquid flow		Cross flow	
Type of trays		Valve	
Type of caps		—	
Vac. design —		Insulation 40 mm	
Code A.S.M.E.			

Nozzles	Mark	Flow rate (kg/h)	Size(mm)
Feed	A; B; C;	11 840	160
O'head vapour	D	23 266	180
Reflux	E	13 106	60
Vapour return	—		
Bottoms	F	1680	60
Drains			
Reboiler vapour	H	30 000	250
Reboiler liquid	I	30 000	150
Draw off	—		
Safety valve	J		50
Manholes – trays	O, 8, 16, 25		500
Thermocouples	K; K;		20
Level gauge	L; L;		50
Pressure gauge	M		20
Level control	Off L;		
Vent	N		50
Steam out	G		50

Remarks: Minimum turn-down ratio 4:1
        see tray data sheet for further details
Information to be furnished by   Project Engineering Division

Construction	Inspection
Stress relieve	Radiograph
Insulation-type	

Issue	Date	Description	App.
1	12.1.72	A-1101; D103	G.L.W.

Figure 8.1 Tower process specification

(m) External cladding and finish.

(n) Estimated cost of vessel.

(o) Final optimization.

To indicate how some of these criteria are applied, consider the Fenske–Underwood short-cut procedure for distillation column design.

(i) *Process design and design criteria.* The basic formula used are

Fenske:

$$N_m = \log \left(\frac{x_{LK}}{x_{HK}}\right)_D \left(\frac{x_{HK}}{x_{LK}}\right)_W \bigg/ \log \alpha_{av}$$

where N_m is the minimum number of plates, x is the mole fraction, HK is the heavy key component, LK is the light key component, D is the overhead product, W is the bottoms product, and α_{av} is α_{LK}/α_{HK}.

Underwood:

$$\sum \frac{(\alpha_i z_i)_F}{(\alpha_i - \theta)} = 1 - q$$

where α is the relative volatility, i is any component, F is the feed, q is the state of the feed, and θ is the Underwood constant.

$$R_m + 1 = \sum \frac{(\alpha_i x_i)_D}{\alpha_i - \theta}$$

where R_m is the minimum reflux ratio, and $\alpha_{LK} > \theta > \alpha_{HK}$.

Bubble-point:

$$P_B^\circ = \frac{P}{\Sigma (\alpha_i x_i)_D}$$

where P_B° is vapour pressure of pure component at dew-point, and P is overall pressure.

To calculate the optimum reflux ratio and plates use the graphical correlations of Winkle and Todd (1971) who have indicated the validity of their technique compared with more rigorous procedures. To use the correlations, Figure 8.2, apply as follows.

(1) From mass balances using the feed rate and composition and the desired recovery and/or composition of the products, calculate the distribution of the components in the distillate and bottoms. Note how 100% purity of a product is not feasible.

(2) Calculate the dew-point temperature of the vapour from the top plate and the bubble-point temperature of the bottoms at the desired operating pressure. Are these temperatures satisfactory? Note that relative volatility goes down as pressure increases and if the reflux ratio is taken as approximately proportional to $1/(\alpha-1)$ then as pressure increases the reflux increases and thus costs. However, usually tower economics are not the main consideration but rather condenser water temperature or thermal degradation in

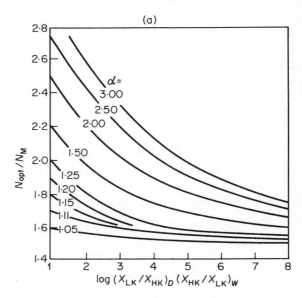

Figure 8.2 Correlations for sizing distillation columns

the reboiler. Refrigeration is expensive and if cooling water is supplied at, say, 23 °C the boiling point of the overhead is limited to, say, 35 °C. Thus as a rule of thumb choose the lowest pressure above atmospheric to give condensation under a normal cooling medium. Avoid vacuum if possible. Similarly try to achieve vaporization with the utility which is most economic.

(3) Calculate α_{LK}/α_{HK} at the average column temperature, or at the feed-plate temperature if it is a 'normal' column, i.e. not a stripping or rectifying column.

(4) Determine the value of N_m from the Fenske equation.

(5) Using the value of $\log\{(x_{LK}/x_{HK})_D{}^*(x_{HK}/x_K)_W\}$ calculate N_{opt} from Figure 8.2(a). This is based on an approximate cost correlation (Winkle and Todd, 1971).

(6) Calculate the value of $\log\{(x_{LK}/x_{HK})_D{}^*(x_{HK}/x_{LK})_W{}^*(x_{LK}/x_{HK})_F{}^{0.55\alpha}\}$ and determine the optimum reflux relationship from Figure 8.2(b).

(7) Determine θ from Figure 8.2(c) using the ratio of $(x_{LK}x_{HK})_F$ and α if the feed is a saturated liquid. If $\alpha < 1.5$, it is recommended that N_m be calculated by the Fenske equation, and θ be checked by $\Sigma\alpha X_F/(\alpha-\theta) = 1-q$. For other conditions of feed we recommend solution using the golden section, Table 5.2.

(8) Substitute for θ and evaluate R_{m+1}.

(9) Calculate R_m and using R_{opt}/R_m determined from Figure 8.2(b) calculate R_{opt}.

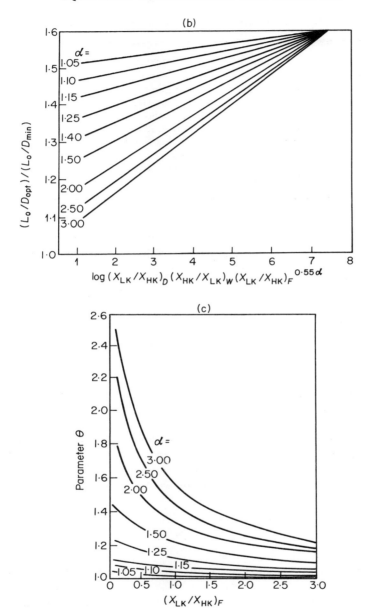

(b)

(c)

(10) Tray selection may now be considered. The range of trays continues to grow and because of the difficulty of selection just generalities will be discussed. Note that 80% plate efficiencies were used in the above correlations. Some common types are listed in Table 8.1. Three basic trays are

Table 8.1. Some absorption and distillation trays and packings

Trays	Packings
bubble caps	Raschig rings
sieve	Lessing rings
ripple	Pall rings
Uniflux	Berl saddles
ballast	Intalox saddles
Kittel	mesh
Glitsch	Glitsch grids
float valves	

Rating criteria, 1 poor to 5 good

	Bubble cap	Sieve	Valve	Counterflow	Pall
Vapour capacity	3	4	4	5	5
Liquid capacity	4	4	4	5	5
Efficiency	3	4	4	5	5
Pressure drop	5	3	5	1	3
Cost	3	4	4	4	2
Flexibility	4	3	5	2	1
Reliability	4	4	3	2	2
Maintenance	3	2	4	3	1

See also Gerster, 1963.

crossflow (with downcomers), counterflow (liquid and gas through the same holes), and packed. The criterion for selection of internals have been tabulated using a scaling method ranging from 1 poor to 5 good, Table 8.1.

Tower diameter is calculated using the Souders and Brown coefficient (1939), where

$$K = 3 \cdot 28 V_s \left(\frac{\rho_v}{\rho_L - \rho_v} \right)^{1/2}$$

where V_s is vapour velocity (m/s), $\rho_v \rho_L$ are vapour and liquid densities (kg/m^3), and K is a constant. The constant K is for bubble cap columns and standard texts give this as 0·18 for 18 inch spacing. The table indicates that valve trays can cope with a higher vapour flow up the column, so if using a valve tray increase K to 0·22. Use average densities over the column to calculate V_s. Clearly the vapour load Q_v depends on an overall heat balance. If sensible heat may be neglected the vapour load is often constant over the column and

$$Q_v = V = (R+1)/D = V_s \rho_v A$$

where D is the distillate flow (kg/h), A is the cross-sectional area (m²), and V is the overhead vapour rate. The ancillaries may now be specified.

Example 8.1. Separation of benzene and toluene in the presence of higher aromatics. 10 160 kg/h of overhead product of 99·96 mole % purity benzene are required. 11 840 kg/h of feed containing 1 mole % of heavies (molecular weight 150) is used. This feed enters as liquid at its bubble-point temperature. The operating pressure is $2 \cdot 3 \times 10^5$ N/m³. Determine N_m, R_m, N_{opt}, R_{opt}. Details are not given of the bubble-point calculations.

(1) Stream compositions:

Component	Feed F (mole %)	Distillate D (mole %)	Bottoms W (mole %)
Benzene (LK)	87·08	99·96	9·20
toluene (HK)	11·75	0·04	82·53
heavies (HK+1)	1·15	0·00	8·27

(2) Temperatures:
$$t_{top} = 95 \text{ °C}; \qquad t_{bottom} = 130 \text{ °C}; \qquad t_{av} = 112 \cdot 5 \text{ °C}$$
(3) Relative volatilities:
$$\alpha_{LK} = 2 \cdot 46; \qquad \alpha_{HK} = 1 \cdot 0; \qquad \alpha_{HK+1} = 0 \cdot 4; \qquad \alpha_{av} = 2 \cdot 46$$
(*N.B.* HK+1 is heavier than HK.)

(4) $\log_{10} \{ (x_{LK}/x_{HK})_D (x_{HK}/x_{LK})_W \} / \log_{10} \alpha_{av}$
$$\log_{10} N_m = 1 \cdot 1$$
$$N_m = 12 \cdot 6$$

(5) From Figure 8.2(a),
$$N_{opt}/N_m = 1 \cdot 96$$
$$N_{opt} = 24 \cdot 7 = 25$$
(6) $\log_{10} \{ (x_{LK}/x_{HK})_D (x_{HK}/x_{LK})_W (x_{LK}/x_{HK})_F{}^{0 \cdot 55 \alpha_{LK}} \} = 5 \cdot 656$

From Figure 8.2(b),
$$R_{opt}/R_m = 1 \cdot 47$$
(7) From Figure 8.2(c),
$$\theta = 1 \cdot 16$$

(8) $R_m + 1 = \sum \dfrac{\alpha_i x_{i,D}}{\alpha_i - \theta} = 1 \cdot 88$

Therefore $R_m = 0 \cdot 88$.

(9) $R_{opt} = (0 \cdot 88)(1 \cdot 47) = 1 \cdot 29$

(10) $K = 3 \cdot 28 V_s \left(\dfrac{\rho_v}{\rho_L - \rho_v} \right)^{1/2}$

For a valve tray $K = 0.22$ for a spacing of 460 mm. Vapour density, $\rho_v = 3.60$ kg/m^3. Liquid density, $\rho_L = 815.4$ kg/m^3, $V_s = 10.12$ m/s

$$\text{Column diameter} = \left(\frac{4Q_v}{\pi V_s \rho_v}\right)^{1/2}$$
$$= \left\{\frac{4(R_0+1)D}{\pi V_s \rho_v}\right\}^{1/2}$$

Column diameter $= 1.50$ m

Based on the information available the specification sheet, Figure 8.1, is completed, nozzle connections sized, and the unit costed.

Opinions differ as to the accuracy of these short-cut procedures; compare Winkle and Todd (1971) and Guerveri (1969). They can be used for location of the starting conditions for plate-to-plate computer evaluation, using the procedures of Thiele and Geddes (1933) and Lewis and Matheson (1932). Of the latter one can only suggest that, whichever method is available, so this should be used. The accuracy of these programs also depends on whether it is possible to optimize the column without using actual operating costs and incorporating heat exchanger design into the system of which the column is a part. To do so requires an extensive plate-to-plate calculation and normally considerable computer time. Typical of the results obtained using the full procedure is Table 8.2 which did not indicate very great savings by the more rigorous method. Of course for other systems this may differ.

Table 8.2. Optimum costs for a distillation system

q values	0	0.3	0.6	1.0
Capital cost of column+pumps	53 799	55 875	55 295	56 386
Capital, E_4, overhead condenser	10 680	10 680	10 680	10 680
Capital, E_3, feed preheater	6 733	5 673	4 336	0
Capital, accumulator	775	775	775	775
Capital, E_2, reboiler	12 512	12 890	13 791	14 483
Operating costs, E_4	13 676	13 676	13 676	13 676
Operating costs, E_3	34 405	24 083	13 761	0
Operating costs, E_2 +pumps	43 262	53 583	63 901	77 663
Total capital (£)	84 499	86 293	84 877	82 324
Total operating costs (£)	91 343	91 342	91 339	91 339
Cost function U (£)	115 003	115 504	114 104	114 390

(ii) *Final process specification.* For preliminary and budget estimates, in most cases little further specification is required. To obtain the equipment costs use overall factorial methods or a budget estimate from a manufacturer. It is useful to indicate preferred materials of construction. If these are shown to be unsuitable for the mechanical design this should be referred back to process engineering with any suggestions.

For a project going firmly ahead, mechanical specification is commenced and the full specification sheet completed. A cost for the control estimate is obtained after bid-tabulation. Unfortunately, if this is the lowest tender, it is generally assumed to be an optimized value, which it is not if process engineering has not done an optimization exercise. But, in turn, process engineering needs this figure before it can optimize. How is this resolved? In many cases by assuming the process engineering calculations automatically give an optimized answer. Chapter 11 endeavours to show what nonsense this is in theory, but do not jump to conclusions. The increased design costs to find the optimum solution may not always be worth while. If the computer procedure producing the information at the correct stage is not available, when should optimization be attempted? No procedure can be recommended for every eventuality, but in general it is best if process engineering obtain costs of equipment as per budget estimation. This is used as recommended in Chapter 11. When accurate information becomes available following full specification and tender estimates, the value produced as the optimum should be checked.

The nozzle connections and piping sizes may be produced at the process, or more usually the mechanical specification, stage of the control estimate. Clearly this must be carried out in conjunction with the engineering flow diagram produced as indicated earlier. Final orientation may be left until manufacturers' drawings become available.

8.2.2 MECHANICAL SPECIFICATION

At the control stage of the project considerably more effort is required. The mechanical design section requires a specification of the item of equipment sufficient for a manufacturer to be able to interpret, to carry out work, and to produce the product. The data supplied depend on the nature of the equipment, similarity with other items, and the familiarity of the manufacturer with the requirements. Thus a pump for a particular duty could be specified by catalogue number. A storage tank, similar to a previous vessel, could be specified by 'make another tank'. Alternatively a reactor of novel design with special catalyst support trays, nozzles for instruments, and complex gas distributions may require several drawings and manufacturing specifications.

Whether such information comes from the client or vendor the data sheet must be completed. This must include a sketch of the item showing desired piping connections and vessel size. Several sheets may be required for a piece of equipment, for instance agitator details might supplement a particular reactor design. The information produced from the engineering flow diagrams must be integrated with the equipment design.

Overall mechanical specifications are written for most items of equipment. They represent the standard to which work will be carried out by the companies' contractors.

Thus the control house, switch house, and amenities buildings will be specified as well as the contents. Similarly, pipework is not only specified by the materials and diameter but the pipe supports, welding procedure, and lubrication procedures are detailed. Testing and inspection methods are listed. Other items include structures, concrete and reinforcing, electrics, drainage, insulation, fire-proofing, and steam tracing.

For items of equipment similar specifications are made. Thus for, say, a centrifugal pump the specification will be covered as follows.

(a) A general specification which covers the requirements for selection and mechanical design of centrifugal pumps in, say, oil refinery service.

(b) A data sheet for each pump which defines process duties and any departure from the general specification.

The general specification covers pump types, materials of construction, casing design, impellor selection, shafts and sleeves, stuffing boxes, bearings and lubrication, auxiliary piping, base plates, couplings, balancing, guarantees, shop-tests, inspection, preparation for shipment and field inspection, welding, and miscellaneous items such as nameplates, tools, and suitability for outdoor duty.

For example, exchangers are designed to codes of practice such as the *Thermal Standards* of T.E.M.A. and the *Mechanical Standards* of T.E.M.A., Class 'R', and *A.S.M.E. Code Section VIII*. Any departure from these specifications must be noted. Thus design pressures are normally the following:

(a) the operating pressure plus 10%,

(b) the operating pressure plus 20 lb/in² (1·4 bar),

(c) 75 lb/in² (5 bar),

whichever the greater (not vacuum duty).

Corrosion allowances should be specified. The codes of practice are particularly useful with regard to the thermal limitation of tubes of different materials. Thus for a specific design temperature and pressure an acceptable tube may be obtained from tables. This does not allow for corrosion properties of the process which should be indicated on the process specification sheet. These should be noted carefully. For example, a superheater in a hydrogen plant was originally designed with Co/Mo tubes to withstand stress corrosion cracking. Slight upgrading of the process conditions resulted in the project engineers selecting a S.S.18 tube based on the tables. The result was that carbonate attack caused tube failure within three days.

In the enquiry stage the client supplies data sheets and any additional specifications applicable in a particular case over and above the requirements of the general specification. In preparing the sheets it is essential to know the range of equipment available. In this respect an index of manufacturers'

Reciprocating

Motor
- Portable
- Stationary
- Metering
- Hand

Diaphragm/bellows

Piston
- General
- Motor
- Metering
- Hand
- Pneumatic

Plunger/ram
- Chemical and industrial
- Metering
- Pneumatic

Rotary

Screw
- Multiple screw
- Single screw

Lobe
- Multiple lobe
- Single orbital lobe
- Helical eccentric lobe

- Sliding shoe

Vane
- Sliding vane
- Flexible vane
- Hand

Gear
- Epicyclic
- External mesh

- Other rotary

Peristaltic

Rotary
- Sinusoidal
- Metering
- Others

- Other positive displacement

Centrifugal

Vertical
- Process and general service
- In-line
- Sump
- High-pressure (100 lb/in²)
- Glandless
- Canned
- Submersible/borehole
- Multi-stage
- Self-priming
- Slurry/unchokeable
- Others
- Glandless
- Close-coupled

Horizontal
- Canned
- Chemical and process fluids
- General service
- High-speed (3000 rev/min)
- High-pressure (1000 lb/in²)
- Multi-stage
- Self-priming
- Unchokeable
- Others
- Borehole
- Others
- Vertical

Mixed flow

Axial flow

Special types and applications

Carboy/barrel
High-temperature
Low-temperature
High viscosity
Hydraulic
Pneumatic
Electromagnetic
Steam driven/engine driven
Jacketed
Variable speed
With lever control/remote control
Portable, mobile
Others (cryogenic, nuclear, etc.) for liquid metal

Reciprocating
Turbo-centrifugal
Cycloidal
Rotary impeller/vane
Rotary piston
Rotary liquid ring
Oil free
Diffusion
Diaphragm

Special materials and alloys

Carbon, graphite
Glass, ceramic
Lined (rubber, ceramic, ebonate resin, plastic, glass)
Lead
Titanium
Bronze
Hastelloy
Plastic
PTFE
Others

Porcelain/graphite steam jet
Steam jet ejector/thermo compressors
Air jet ejector
Mercury vapour jet ejector
Liquid/water jet ejector

catalogues is invaluable. Table 8.3 is taken from the Chemical Engineering Product Data file of Technical Indexes Limited and indicates the range of pumps contained in their files. Each pump type is referenced by a list of manufacturers and their catalogues.

The vendor in the tendering stage carries out mechanical design (or process design if deemed necessary) of the unit and makes such calculations as are necessary, so that in the event of an order the design meets the requirements of the relevant statutory bodies and the vendor can accept full responsibility for the unit. Several tenders are obtained from different vendors and a bid-tabulation prepared. These give details of aspects of the design, the capital cost plus any extras, delivery dates, etc. If the costs differ from the preliminary estimates it may be necessary to repeat some optimization exercises. Otherwise the bid-tabulation is analysed using the procedure of Example 4.7. Sometimes it is cheaper for the vendor to supply a larger unit than requested by standardization with his range. Alternatively he may supply, say, an exchanger in two equal sizes which combined are equivalent to the original design. In this case the installed cost of units should be compared. This approach is, of course, streamlined when providing a full manufacturing specification or manufacturing the units oneself.

8.3 Integrated equipment design

The ideal approach is integrated equipment design. The principles are indicated for pressure vessels. It is desirable to have a program which performs all the process and mechanical design calculations associated with pressure vessels. This in turn should be a subroutine of an overall process simulator (Chapter 12). Ideally such a program would be carried out using a computer graphics facility. Such a facility is indicated in Figure 8.3. The facilities provided for the user by the program are accessed by means of light buttons and function keys. The light buttons are a set of symbols generated as a menu on the screen. These, when detected by the light pen, execute some programmed action. Similarly the function keys execute a programmed action but these tend to have a common task for different graphics programs and are basic draughting controls. The computer requires a file of data and backing stores such as disc and magnetic tape are required.

The procedure for pressure vessel design is indicated in Figure 8.4.

(i) *Initialization mode.* The initialization input to the program includes the facility to select from file equipment data and specification sheets.

(ii) *Process specification mode.* In the process design section information must be input to the program on design pressures and temperatures and process materials. The volume of the vessel may then be calculated. This information may be based on information from sales and despatch, depend on analytical schedules, a full process design, or simple rule of thumb.

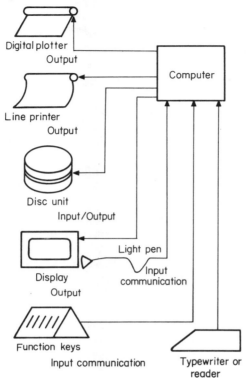

Figure 8.3 Computing configuration

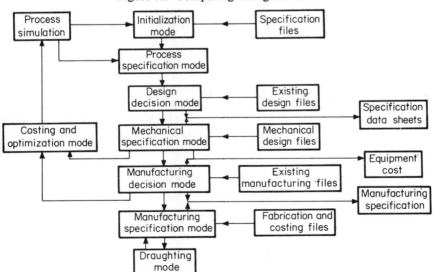

Figure 8.4 Structural diagram for pressure vessel design

Examples of the latter are for liquid hold-up such as in steam drums allow 15 minutes' storage of boiler feed water to prevent loss of water level; or for reflux accumulators choose whichever the greater of 2 minutes' product to storage, 15 minutes' product to subsequent tower, or 5 minutes on reflux. The liquid holding volume of a surge drum might be 50% of the vessel or, for a storage tank, 80%. It is convenient to enter the design code at this stage. These might comprise B.S. 1500, B.S. 1515, A.S.M.E. VIII, and any special company codes. The material required for fabrication and the corrosion allowance should be entered, particularly if corrosive conditions exist.

(iii) *Design decision mode.* Based on this information the design decision mode is entered. What type of vessel is to be used, i.e. standard, spheres, floating roof, bullets, etc., for storage tanks? From the file of existing designs, including external supply, can one be selected and, if so, is it to be modified or used in creating another? Any useful information on file is carried forward to the next mode.

(iv) *Mechanical specification mode.* We now enter the mechanical specification mode and many designs enter the program at this stage only, the earlier decisions being made using normal design procedures and the specification data sheet being filled in manually.

Diameter: this may be selected on the basis of a length to diameter ratio which has in general proved to be in the economic range of $L:D$ ratio. It may be influenced by minimum instrumentation dimensions which set constraints or in exceptional cases by layout. Preferably an optimization procedure should be written. This may involve several iterations through the program before selecting the optimum design.

Shell thickness: calculated using stresses given in the appropriate codes. Stiffeners are incorporated as required and an iterative procedure is used. This will specify the material if not previously decreed. Otherwise the data are fed in manually from the codes. Thus, for shells, the design stresses at design and room temperatures must be input.

Ends: these are designed as per appropriate code. They include dished, hemispherical, bolted, welded, and conical. The appropriate type is selected by the user and the design information inserted as required.

Nozzles: from the engineering flow diagrams the number of flanges is inserted. Each connection is then designed. Further allowance is made for any compensation and type of support. It is desirable to calculate the weight of the vessel. This information is printed together with a sketch of the vessel on the specification data sheet.

(v) *Manufacturing design mode.* A decision is now made with respect to manufacturing procedure. The specification sheets may be sent out to specialist manufacturers and individual tenders obtained or comprehensive information provided to permit the manufacture of the unit. Files and technical indexes assist in making this decision.

Table 8.4. Cost analysis of pressure vessel

Reference, F–806, cyclohexane plant

Shell	Diameter	Thickness	Weight	Rate	Amount	Cost
Side	1500	16	1600	0·100	160	
End 1	1500	30		dished	100	
End 2	1500	30		dished	100	360
Nozzles:						
1	600	10		150	150	
1	100	10		25	25	
3	80	10		20	60	
2	25	10		10	20	255
Blanks					35	
Instrumentation					0	
Manholes 1	500	16		104	104	
Saddles			300	0·200	60	
Lining (m)					0	
Painting (20 m^2)				0·4	8	
Radiography (10 m)				4	40	
Surface preparation (20 m^2)				0·4	8	
Stress relief (3000 kg)				0·01	30	
Total miscellaneous						285
Basic equipment cost (£)						900
Overhead, drawings, delivery, and profit at 20% (£)						140
Total cost of vessel (£)						1040

(vi) *Manufacturing mode.* Detailed mechanical design as well as incorporating codes of practice must include a study of the availability of materials and labour costs as indicated below in the costing and optimization mode.

(vii) *Draughting mode.* The sketch of the vessel may be upgraded using the draughting facilities of computer graphics. A library of shapes should cover standard flanges, typical branches, shell and end details for specific shapes, standard covers, davit arms, and miscellaneous internals. To access a basic shape requires the name of the type and a list of sizes. These shapes are supplemented with the ability to draw lines, arcs, and circles and to delete. The picture must be fully manipulable, using the light pen in conjunction with the light buttons and include the facility to expand, to contract, and to scissor the picture on display. Captions and dimensioning is required. Finally the drawing is printed together with the specification. As well as being available for manufacture the data are also stored for access at the design decision mode on future projects.

Table 8.5. Percentage factors for installation of main plant items taken as 100% of cost of individual main items

Value of individual main items	Over £20 000	£10 000 to £20 000	£5000 to £10 000	£2000 to £5000	Under £2000
Erection					
Some site fabrication	9	10	12	14	19
Average	5	6	8	10	15
Erection included in cost	1	2	3	6	8
Civil					
Sitting on floor	1	1	1	2	3
Average	5	6	7	11	20
Piping					
Liquids and gases or complex liquids	16	24	30	60	90
Liquids in average system	12	18	24	50	60
Small bore	5	9	13	25	40
Ducting and chutes/solids	2	4	6	10	20
Insulation					
Average	2	3	5	15	17
Service only	1	2	3	5	10
Instrumentation					
Local	2	3	5	10	20
Three controllers and ancillaries	15	26	35	48	70
Two controllers and ancillaries	10	16	26	36	50
One controller and ancillaries	7	10	18	27	40
Buildings and structures					
Open air plant with minor building	6	8	10	11	30
Open air plant within structure	12	20	22	30	45
Plant in simple covered building	15	24	30	40	60
Electric					
Lighting and power for ancillary drives	8	9	11	16	30
Lighting and power for m/c main drives (excluding switchgear)	10	12	14	20	35
Lighting and power for m/c main drives (including transformers and switchgear)	15	18	20	27	50

Notes.—The table was prepared from a survey of costs mainly from *Capital Cost Estimation* (1969) and Miller (1963). The costs are 1972 figures. The survey gave poor correlations of factors from different sources for large-bore piping in complex systems, erection of items involving much site fabrication, extensive lagging, plant in a major building electrical for electrolytic plants, and heavy foundations. These costs all involve high factors and should be calculated separately using appropriate design procedure. Overheads should be added to the compiled total cost. For plants costing over £50 000 add a further 25% in the absence of definite costs.

If using the factors for optimization ignore any aspect which is the same for all alternative units, i.e. instrumentation on columns, but ensure these are added in for the final estimate.

(viii) *Costing and optimization mode.* This costing mode necessitates a detailed schedule of rates for building up the equipment cost. A typical build-up based on mythical values is given in Table 8.4. Obviously this will apply to only very common materials and special corrections are required for different materials and upgrading cost information. It is rather impracticable for truly iterative calculations to be used for optimization and normally the user inserts alternative input information as he considers necessary. This cost is only approximate and developed for optimization purposes. Real costs are influenced by availability of dished ends, standard plate width, intermediate seams, and welding quality. Overheads and profit are very dependent on business considerations of the vendor. Thus, after obtaining a price from a manufacturer, it is necessary to check with the estimates to ensure an incorrect optimization was not made. Of course, if the manufacturing mode is entered a more accurate cost is obtained. For optimization costing should not stop at equipment cost but by means of factors, Table 8.5, should be further refined to installed equipment and combined in the process estimate costs. Care should be taken over the interpretation of the results and if necessary process input data should be modified. All optimization decisions should not violate Bellman's principle of optimality.

Furman and Cheers (1971) have indicated a suitable algorithm for carrying out an exercise similar to that outlined above. It could with ease be built into a computer graphics routine.

General. It should be appreciated that such a plan is for the future. At present only very large companies can develop their programs in such a manner. Closer to reality is individual programs being developed to produce partially completed data sheets, the remaining information being completed by hand. The most important aspect from the point of view of the process engineer is to develop programs which permit going from process data to optimized cost of the unit, at the same time producing a specification sheet to supply a manufacturer for his tender. It is considered that this will greatly increase the role of the process engineer and facilitate his achieving the objective function.

9 Brief Notes on Modifying Considerations of Costs

9.1 General

The modifying considerations were identified under the schematic chart. They essentially modify aspects of the design and adjust the capital and operating costs.

9.2 Site

The site will affect the capital cost of project. As well as the geological state of the ground, aspects such as previous dumping of chemicals, old sewers, etc., can increase the cost of piling and back-fill. Very rarely does the actual cost of the site act as a restricting feature on the development. This is due to the lower cost of the type of land developed compared with the large investment to go on the site. Special site hazards can apply such as abnormal meteorological conditions, flooding, and earthquakes. Is water freely available, if not exchangers may be more expensive? Are transport communications adequate? Usually if the site is developed inside an existing works the cost of site clearance may be higher but the savings in roads, utilities, etc., more than compensate. Even so, road and railway improvements may be required. For information on site layout see the *Report of the Working Party on Chemical Plant Layout* (1973).

9.3 Environmental

With the present emphasis on pollution it is clear that, although in some cases legal restrictions may not impose a limitation, public opinion can. The process engineer is basically faced with two cases.

The new plant—it may not be practicable to install a new plant on an old site because of the increased control of effluents. Alternatively, improved methods will have to be used on the process to clean up the streams. These presumably cost money, must be allowed for in the revised profitability, and built into the basic design.

The old plant—such a plant may not be economic to modify. Here it may be necessary to attempt to blend, say, effluent with a large-volume cooling water stream to dilute the pollution. The process engineer is faced with the moral question which can in extreme cases be summed up by either keeping people in employment at the works in question and living in the area, or for instance having fresh-water fishing and boating. The problem is not as easily resolved as current fashions would dictate. The process engineer requires a knowledge of the legislation, and the problem is integrated in the design. There can be no doubt that basically it is a good trend to emphasize environmental considerations, but this should incorporate a sense of proportion and economic reality.

When laying out plant, try to blend in the background and allow for landscaping. Lay on good amenities for the operators such as adequate toilets, tea-mashing equipment, and a mess room. The profitability of such actions may seem low and difficult to justify but motivation must pay in the long term, particularly if the commissioning is long and arduous.

9.4 Plant layout

Layout represents a situation in which compromises are made by a mixture of good practice, opinion, experience, and economics. The theoretical minimum space a plant can occupy is the total volume of its components. Various constraints prevent the attainment of this minimum; such constraints include allowing adequate clearances for access during operation, maintenance, and construction (allow for installation of late-delivery items) and to allow safe operation. The unit plot plan is produced after a minimum cost exercise to optimize pipework, structures, drains, etc., within the above constraints. Elevation is rarely a constraint on a layout. It is usually only necessary when, by utilizing gravity, economies can be made in capital, site space, and operating costs. Only occasionally does a genuine case exist where pumping of slurry, etc., is impossible. Some typical economies by elevation arise from using thermo-syphon reboilers and providing a net positive suction head for centrifugal pumps. Stacking of heat exchangers, etc., can give savings and economic formulae have been derived (Bush and Wells, 1971). Unlike pipework, every decision is a special case depending on the structures, pipe-bridge, weights, pipework, etc. Consequently, rules of thumb are useful and in general the decision is to stack if the engineer feels that he will thereby achieve a layout better in terms of appearance and operability. However, if maintenance problems are anticipated stacking should be avoided.

A study of layout can indicate the necessity to affect compromises with good practice. This principle affects other aspects of process engineering. Assuming the objective to be optimized is economic; if good practice decrees an excessive time on design, thereby increasing costs, or alternatively creates delays in which rivals may bring out processes or obtain sales contracts, then good practice itself will vary.

9.5 Safety

Safety is a function of many aspects largely within the control of the process engineer. Good design of engineering flow diagrams and reactors is vital, as too is avoiding compromises with good layout. Read up on other people's accidents to ensure you are doing everything you can. Of vital importance is good operator training. Ensure valves are not opened to divert a stream unless the line and its outlets are fully checked. These aspects do not cost a great deal whereas down-stream time does.

If the plant does have a fire, make sure that it is roughly known who is going to tackle the fire and who is going to fail-safe the process. Ensure that proper fire-fighting facilities and capability of obtaining help is provided. It is important to consult early all authorities who can advise, such as factory inspectors, fire prevention officers, and works safety officers.

9.6 Maintenance

Allow adequate facilities and access for maintenance at the layout stage. The provision of spares, etc., is a function of the initial capital cost of the process. Whether to install stand-by pumps is dependent on the reliability of the process. However, according to the reliability of the process so maintenance becomes involved, whether or not this be scheduled. The two schools of thought regarding maintenance organization are shown in the maintenance of pumps with stand-by. These may be either overhauled when they break down or taken off-stream at regular intervals for servicing. Either way the plant has problems if during this time the on-stream unit fails.

However, as far as maintenance is concerned the process engineer will mainly be involved with the cost allocated in the profitability. This has been discussed under balance sheets. It is important that it be clear whether or not the value allows for maintenance overheads such as partial cost of machine shop, etc.

9.7 Utilities

Utilities are required for the supply of steam, plant air, compressed air, fresh water, boiler feed water, distilled water, cooling water, electricity, provision of inert atmosphere, fuel gas, fuel oil, etc. Alternatively they may be a by-product of the process. Preferred sources of energy may be specified, i.e. water not available for cooling duties.

All utility consumptions should be measured and charged to the plant. Aspects such as heat for operators' mess room and warehouses should always be allowed for in proposed new plants and credited against the process. Usually if, say, steam is produced and consumed on the same plant it tends to be unrecorded as a cost. This practice is very inefficient with regard to cost control. Always ensure that P.F.D. and E.F.D. are provided for design of utilities and an operating manual produced. This should include a full specification for each of the ultilities denoting for instance the difference between plant and instrument air.

9.8 Pipework

Obviously the cost of pipework is a major part of the capital cost of the plant, and according to the material of construction, so the effect of this modifying consideration will increase. At the control stage the cost is usually based on a probable extent of supply with rates for variation. An important task of the process engineer is to ensure the use of correct materials process-wise, accurate optimization procedures, adherence to engineering flow diagram and that the pipework when it is handed over is clean and pressure-tight. Pipeline sizing is discussed in the section on optimization.

9.9 Control

Two aspects of instrumentation are important with regard to operation.
    (a) It reduces labour requirements.
    (b) It improves performance.
As the cost can vary considerably according to the degree of instrumentation, expertise is necessary in attempting to allow for instrumentation costs at an early stage of the project. Normally it is best to make full instrument schedules.

The difficulty of assessing such cost is shown by one job a major contractor carried out abroad. The cost of labour was so cheap that rather than spend money transmitting signals to the control room a man could be left to watch the gauge and to maintain the level by hand.

Owing to insistence on cost control and adhering to budgets on some projects, because instrumentation is one of the last items to be purchased considerable pressure is made to reduce expenditure. This is difficult to withstand because much of the savings arise from difficult-to-determine aspects, such as reducing the time of commissioning and improving plant performance.

9.10 Operation

Management and labour requirements vary according to the different plants. At commissioning they are amplified by process engineering assistance. Work measurement can assist in defining the requirements as can assessments based on experience on existing plants. Full allowance must be made for non-working time operatives. When instrumenting chemical plant it is as well to leave some readings out on the plant so that the operators do not become too office-bound. After all, forcing an operator to walk round the process using all his senses can give an early warning of trouble (even the approach of the works manager).

Operating manuals vary in complexity. Personally I believe in providing excess of information. A list of contents is as follows.

1. Basis of design
   1.1. General design specifications
        1.1.1. Capacity and degree of flexibility
        1.1.2. Feed specifications
        1.1.3. Product specifications
        1.1.4. Process design criteria. The main parameters relating to the design equipment
        1.1.5. Definitions—define phrases that might be subject to more than one interpretation, i.e. yield, fractional conversion, regeneration efficiency
   1.2. Summary of catalyst requirements
   1.3. Interconnection with other units with particular reference to terminal points
   1.4. Description of process flow
   1.5. Control of performance
        1.5.1. Operating variables
        1.5.2. Analytical requirements
   1.6. Utilities—consumption and production
   1.7. Process flow diagram

2. Process details
   2.1. Control of operation. Description of main instrumentation and control
        2.2.1. Initial start-up
        2.2.2. Normal start-up
        2.2.3. Normal operation
        2.3.1. Shut-down general
        2.3.2. Normal shut-down
        2.3.3. Emergency shut-down
   2.4. Engineering flow diagrams
   2.5. Equipment data sheets and schedules
   2.6. Instrument schedules
   2.7. Brochures

Part 3

THE OVERALL PROCESS, OPTIMIZATION, AND SIMULATION

Chapters 10–12

We are now considering the overall process. Synthesis which was considered earlier should be supplemented with techniques for comparing different systems. Two important considerations are (1) that all decisions should constitute an optimal policy with regard to the state resulting from the first decision, and (2) that the system under consideration be identified as part of a larger system.

This section emphasizes the study of the overall process when optimizing. Dynamic programming is difficult to implement for real systems but it shows the importance of the first principle. Linear programming is readily applied and in wide use.

Finally, simulation is discussed. Because the systems under consideration are large, this is invariably associated with computer-aided design. This approach is strongly influenced by the availability of programs, so only a brief review is presented.

10 Techniques for the Screening of Alternatives

Several techniques have been evolved for the roughed-out optimization of plant route and items. Such techniques harness the creative ability of the process engineer. They range from involved group rituals, useful for generating ideas for later study, to creative logic as in critical examination.

Rudd and Watson (1968) recommend a preliminary screening of ideas during which the engineer probes the concept, asking the following questions.

(a) Is the concept illogical or clearly inferior to an alternative (see also Table 6.3)?

(b) Is the concept inferior to a known inferior processing concept?

(c) Is too great a degree of technical extrapolation required? Some companies have had major start-up problems on the early, 'very large' plants.

(d) Is the concept unsafe or does it suggest a better alternative?

(e) Is competence additional to the company's own expertise required? That is, if you only sell chips, you are unlikely to open a fat refinery. Such questions assist in the initial generation of the flow sheet.

Assuming the concept to be valid, critical examination uses a standard format, to ask what, how, when, where and who? See Table 10.1 for a modified version.

The questions can produce different answers according to the design stage of the project. Elliott and Owen (1968) have indicated its use in the chemical environment and the reference details this approach.

A similar set of questions are produced from the logic of systems engineering (Rippin, 1969). Any system must have an objective. To set the problem in its proper perspective, after establishing both problem and system, the engineer asks the following questions.

(a) What is the objective of the system?

(b) What is the system of which this system forms part?

(c) What is a relevant economic criterion for measuring the efficiency with which the system can achieve its objective?

The criterion may be improved by experimentation or model-building and simulation. As with other questionnaires, different detail is required at

117

Table 10.1. Modified critical examination sheet

What is the system?	The present facts	Proposed alternatives	Selection for development
what is the objective function? (usually economic)	why?	what else influences this system?	what should be the economic criteria?
what is achieved? (sentence in minimum detail)	why? (ask each part of sentence)	what else could be achieved? (eliminate modify avoid other concepts)	what should be achieved? (short term long term economic criteria)
how is it achieved? (outline method; sequence and activities)	why that way?	how else could it be achieved? (materials can be changed extra or different method altered change equipment specification optimize)	how should it be achieved? (select items best fulfilling objective function)

'When?, where?, and who?' may also be asked.

various stages of the project. The key factor is never to omit the objective function from the considerations.

This approach is valid at all stages of the project. It is used in developing the original research, process development, analysing equipment, plant optimization, etc. First let us illustrate the importance of generating alternative action.

Example 10.1. Nitrobenzene plant. In this case the plant, although old, is still functioning and the company concerned had requested the technological department to investigate the profitability. The process involves reacting benzene and nitric acid in the presence of sulphuric acid. Both benzene and sulphuric acid are produced on the works. The sulphuric acid is diluted with water and contaminated with aromatics. It is recovered by a concentration process of somewhat great antiquity, expensive to operate, and a key place for improvement.

The objective is clearly to maximize profit and the plants concerned include the benzole, sulphuric acid, and nitrobenzene plants. Should the nitrobenzene plant be closed? Can benzene and sulphuric acid be sold elsewhere? Are transfer prices correct? The overall system must be studied.

With respect to the acid concentration unit, clearly we may improve the existing process or install a new process. However, a third approach is adopted; sell the dilute acid at a reduced cost. Although this means a slight loss of revenue to the sulphuric acid plant due to replacing some higher priced sales, there is a considerable overall gain to the company due to avoiding operating costs.

Not a great deal of technical knowledge was required to solve this problem; just an ability to generate alternatives.

The value of the techniques may be seen by attempting more formal analysis of specific problems. The critical examination sheet may be used to check the process flow diagram, Figure 6.13. Study the section from the reactor outlet through the heat exchanger sequence to the separator. 'What is achieved?' is the separation of gas from liquid. The efficiency of heat exchange is important but not the key operation. Probing this sentence can suggest the reduction of the hydrogen content of this stream with possible economies. In an earlier case for a similar process, hydrocracking creosote, such action reduced the ratio of hydrogen:creosote from 20:1 to 3:1, carbon formation in the reactor being the limiting consideration. The heat exchanger aspect emerges later when asking 'how is it achieved?'. Reaction products are passed at various temperatures and pressure through reboilers and heat exchangers to a separator. Analysis of this aspect results in a comparison with producing steam in a waste heat boiler or further optimizing the heating sequence. Would the reliability of the process be increased if additional storage were provided after the separator and steam used to reboil the distillation train? Should the separator be operated at a pressure closer to

Table 10.2. Critical examination of reaction diagram, Figure 6.2

What is the objective?	Why?	Proposed alternatives	Discussion of alternatives	Action
to improve the realization of crude benzole by a hydro-cracking process	1.1 to improve profitability	eliminate coke ovens	guaranteed output for next 10 years means this unlikely. Could be partial closures	incorporate possible loss of raw materials in D.C.F. analysis of new plant
		leave benzole in gas	sale of gas a problem. Also burning characteristic affected. Extraction plant in good condition and optimized. In foreseeable future the value of crude benzole as determined by outside sales means it is economic to extract	extract from coke oven gas
		sell crude	present course of action. Selling price fixes cost of raw materials to new plant	
	1.2 why a hydrocracking process?	join with another company in similar enterprise. Six alternative processes being investigated	alternative to present action being pursued (see report 68069)	economic balance economic analysis to be carried out
		eliminate: xylenes	see above, yield of xylenes	before sizing

	eliminate: carbon disulphide	if material transported it is difficult to sell all products and creates an effluent problem. Dicyclopentadiene plant not economic. Some defronting will still be required on own production	eliminate dicyclopentadiene plant. Defront crude at producers
	eliminate: tar bases	as heavies despatched separate economic case made on tar bases plant. Favourable. Export to existing plant at an alternative site not economic	incorporate as tar bases plant on unit
	modify: treat petroleum benzoles	economic study indicates at present purchase price too high. Technically feasible. Not economic to batch through process. Preferably added after prefractionator	sales to keep position up to date
1.3 why hydrogen: required as chemical constituent for dealkylation and desulphurization	eliminate: other processes available	consider these other processes as in this example	consider other economic cases
	eliminate: use other source of hydrogen	electrolytic hydrogen too expensive despite possible plant savings. Other sites might have hydrogen available. Coke oven gas higher in inerts and high compression cost. Reject on economics	survey of alternative sites

Table 10.2 (continued)

What is the objective?	Why?	Proposed alternatives	Discussion of alternatives	Action
	1.4 why impurities: part of reformer product hydrogen	avoid: remove to low levels of CO and CO_2 by methanation, after CO_2 removal by absorption. All inerts could be removed by regenerator/ cryogenic unit	CO_2 removal cheapest by absorption. Methanation increases amount of inerts but will occur in reaction. Not possible to make decision yet except inerts generally undesirable	reconsider after reaction and process stages considered. CO_2 absorption, cryogenic gas removal and methanation noted
Stage 2 reaction crude benzole is dealkylated, desulphurized, hydrogenated, hydrocracked, isomerized, polymerized, and cracked, in the presence of hydrogen and inerts	2.1 why dealkylate: higher aromatics to benzene. Toluene and xylenes are converted to toluene and benzene	eliminate: would reduce yield of benzene	advantage of process is increased flexibility with regard to B.T.X. fraction and improved yields of benzene. Alternative process only real solution	ask sales how important is increased flexibility. Study details later
		modify: dealkylate only higher aromatics	this would only result in additional toluene and xylene production. Major advantages of this process is in optimizing benzene. If market for xylenes alternative process preferable	reject
	2.2 why desulphurize: to produce sulphur free products	modify: desulphurize prior to reaction	carbon disulphide can be removed at producers, save on plant and hydrogen	accepted earlier

	modify: do not de-sulphurize to above extent	acid washing, etc., of crude highly expensive and technically difficult. Acid washing of benzene, etc., after reaction is possible	reject on economic grounds
2.3 why isomerize to increase p-xylene formation. Also produces diphenyl	eliminate: sell low-quality products	sale only for sulphur free products	reject
	eliminate: increase yield of diphenyl	reaction unimportant as no sales, select catalyst accordingly	reject
	modify: increase isomerization	if xylene yield were higher could change importance owing to higher market value of p-xylene	further development on catalysts?
2.4 why hydrocrack: to remove mainly aliphatics	eliminate: sell low-quality products	not economic	
	modify: remove aliphatics by either means	implies other process	
	eliminate: unavoidable product reduced as much as possible by choice of catalyst and reaction temperature	implies carrying out further research development	ensure operating instructions deal with these facts

Table 10.2 (continued)

What is the objective?	Why?	Proposed alternatives	Discussion of alternatives	Action
	2.5 why crack: produces carbon as undesirable product	avoid: burn off catalyst carbon continuously	this does not avoid loss of product but avoids shut-down, etc. Implies more expensive plant than fixed bed. Study in method	review
		modify: has been claimed that carbon formation builds up to maximum so leave carbon where it forms	theory largely disproved; probably is a slight effect	reject
		remove heavies and have excess hydrogen	this would reduce rate of carbon formation. Optimization required of hydrogen excess. Presence of heavies can have undesirable polymerization effects in reactor	incorporate as in stage 1
	2.6 why hydrogenate: saturates double bond	eliminate: reaction optimized at lower temperatures than necessary above	produces undesirable by-products, if not carried out. Also excess cracking. Pre-treatment may be necessary	incorporate in reaction stage. Cost up separately
		modify: remove tar bases	reduces above requirement but not complete solution. Tar base plant incorporated on economic grounds	tar base plant agreed above

2.7 why polymerization: unavoidable consequence of heating crude and reaction to produce diphenyl	eliminate: research advise not readily possible	further work desirable but unlikely to be promising	review
	modify: remove polymerized products by gravity separation. Recycle of diphenyl reduces amount of benzene which polymerizes	pretreat reactor would assist. Polymerized products separate from crude benzole fairly easily	review pretreat stage when method considered
2.8 why inerts: in product from reformer	eliminate: CO and CO_2 converted to methane in reactor. Consumes hydrogen and generates heat	undesirable side reaction. Reduce to acceptable limit	incorporate carbon dioxide removal
	carbon monoxide poisons catalyst	must reduce to acceptable limit	incorporate carbon monoxide removal
	avoid: select catalyst not poisoned by carbon monoxide	implies further research/ development	review

The further development is left to the reader.

that of the stabilizer? If so, pipework would be cheaper, but does this represent the only change in the economics? Is the proposed sequence of heat exchangers correct? The diagram should be examined and the questions postulated and answered.

Example 10.2. Consider the crude benzole refining process outlined in Figure 6.2. An analysis as carried out by students with only slight knowledge of the process is given in Table 10.2.

Only the first two stages are indicated. The following stages ask the following

---

What is achieved?

Stage 3   Gas/liquid are separated
Stage 4   Liquid products are separated from impurities
Stage 5   Benzene No. 3, toluene-nitration grade, ethyl benzene: xylene are produced
Stage 6   Process impurities are removed from gas for recycle

---

The validity of the technique is shown by the answer previously given in Figure 6.2. Obviously it is not necessary to write down answers in detail, but adopting the formal procedure gives experience in questioning proposals. It is a vital tool in the synthesis of chemical plant.

11 Optimization of Large Systems

11.1 Principle of optimality and dynamic programming

11.1.1 GENERAL

We have considered some of the many optimization procedures available in evaluating the economic criteria. Nearly all were concerned with being given some data and optimizing the values so generated. However, in using the techniques for the screening of alternatives we ask what is the system of which this system forms part? If these data are not part of an optimized overall process then we have not optimized the system.

Example 11.1. Consider a simple process sequence first illustrated by Mitten and Nemhauser (1963).

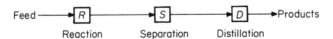

For any given feed operate the reactor in an R_i manner, the separator in an S_{ij} manner to give a product ij and the column in a D_{ijk} manner giving a system product ijk.

Take representative values so that reactor has three modes of operation and the separator and the column have two modes. Assume identical feed and products.

On the network, Figure 11.1, the 12 routes and the cost of each operation are indicated. Various schemes can be analysed directly from the network.

Scheme 1. Add up all 18 routes and pick the cheapest.
        Route $R_1 \rightarrow S_{13} \rightarrow D_{132} = 9$ units

Scheme 2. Optimize R, S, and D separately to obtain $4+1+2 = 7$ units; an impossible route.

Scheme 3. Choose the best first stage; R_2—4 units.
        Choose the best of second stage possible routes; S_{22}—2 units.
        Choose the best of third stage possible routes; D_{222}—5 units.
           Total 11 units—clearly not the minimum.

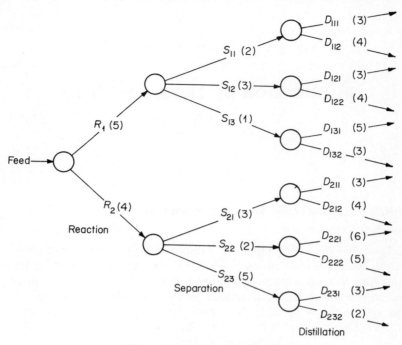

Figure 11.1 Process routes, Example 11.1

Scheme 4. Select optimal distillation route to each node, in distillation section,

$D_{111}, D_{121}, D_{132}, D_{211}, D_{222}, D_{232}$

Add in cost of separation and select optimum route to separation node,

$S_{13} + D_{132}; S_{21} + D_{211}$

Add in cost of reaction and select optimum route,

$R_1 + S_{13} + D_{132} = 9$ units

Scheme 4 was carried out in accordance with the following important theorem.

Bellman's (1957) principle of optimality. An optimal policy has the property that, whatever the initial state and the initial decision are, the remaining decisions must constitute an optimal policy with regard to the state resulting from the first decision.

11.1.2 APPLICATION OF THE THEOREM—DYNAMIC PROGRAMMING

Compare the average process design. Usually the reaction stage is designed first. This is optimized and the design proceeds forward in stages.

Unfortunately, to carry out a full analysis of the network requires storing a considerable amount of material in the computer. In the brief example 20 separate systems have been designed, compared with, say, 5 which might normally have been carried out. Furthermore, each route has to be searched with the utmost care. Rudd and Watson (1968) have simplified the technique by employing straight-chain dynamic programming. They calculate the optimum values for each item over a range, and assume a smooth curve may be drawn through these to generate additional values.

Thus consider a single series of actions. For convenience, consider 'value' as referring to a cash figure and 'state' as an item of data of the system, such as percentage conversion.

A, B, C, and D are cash 'values' and a specific 'state' acts as the link. Optimum 'value' A is an optimal policy to carry out for a specific input 'state' b. If we input a range of 'state' b we can compute an optimum 'value' A in each case, and thus produce a range of these optimum 'values'. Thus we have

Optimum 'value' B is an optimum policy to carry out for a specific input 'state' c. It is selected as the optimum from a range compiled from a 'value' of B and the appropriate optimum 'value' A as decreed by the input 'state' b. A range of optimum B's is produced for various values of input 'state c. Thus we have

followed by

11.1.3 REVERSAL OF INFORMATION FLOW

This work has been done working from the end of the process contrary to the mass flow. However, decisions are based on information, and information flow may frequently be reversed. Thus, by using the principle of reversal of information flow, it is possible to make decisions on optimality going in the same direction as the mass flow. For instance, if we reverse the flow of information on the distillation, separator, reactor sequence, we have Figure 11.2.

The decisions are still optimal although in a contrary direction to the mass flow. Thus, if we wish we can optimize in the reverse sequence to this,

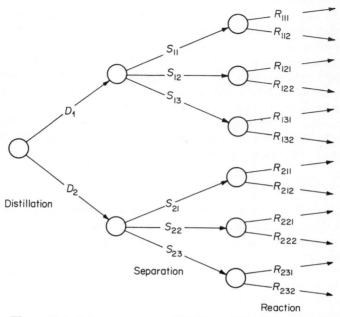

Figure 11.2 Process routes modified by reversal of information flow

reactor, separator, column, which may be considered a more satisfactory manner of working. Let us analyse this in terms of design time. The end of the process at which to start is either the end at which least diversion is likely to occur, or the end at which cost information is liable to be altered. Thus, if the sales revenue is liable to change, then preferably this information should enter last in the exercise. Similarly, if several sales figures are to be used for the same component value, this should be submitted at the end of the calculation to prevent excessive calculation.

Example 11.2. Dynamic programming. To illustrate this principle and dynamic programming the example of Mitten and Nemhauser (1963) has been recalculated, using the technique of Rudd and Watson (1968), except that unlike the above we have reversed information flow. This means that our calculation goes in the same direction as the mass flow. Only outline details of the problem are repeated and further reference to Rudd and Watson (1968) is recommended.

In this problem a fixed quantity of material is processed to a variable amount of product, the remaining material being disposed of as waste. A sketch of the planned process is given in Figure 11.3. Data on the equipment for a range of 'states' are given in Tables 11.1 to 11.8. Note particularly the information flow diagram in Figure 11.3, as this shows the input 'state' into the next stage of the process.

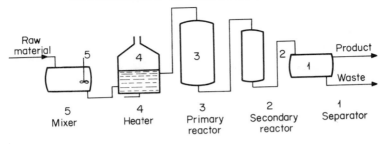

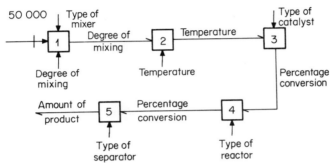

Figure 11.3 Dynamic programming example (excerpted by special permission from *Chemical Engineering*, 20th Sept., 1971. Copyright (©) 1971, by McGraw-Hill, Inc., New York.)

The system is to be optimized for maximum venture profit by the equation

$$V = 0{\cdot}4(S - C - 0{\cdot}2I)$$

where S is the sales income ($/annum), C is the manufacturing costs ($/annum), and I is the investment ($).

For convenience, instead of showing many negative profit values the optimum 'values' are indicated whether they are costs or profits.

Optimum scheme. Tables 11.9–11.13 trace the path of the design network. The optimum appears to be at 60% conversion after the clean-up reactor. A quick check shows that 50% and 70% conversion values are below this optimum value. Thus the optimum scheme is as follows.

(5) 2 separators.

(4) No clean-up reactor.

(3) Primary reactor at 800 °F using catalyst A_1.

(2) Special heater.

(1) Mixer B.

This is identical with the Mitten and Nemhauser design. Furthermore, if the sales revenue changed it is extremely easy to modify the design conclusions.

Table 11.1. Data on the mixing operation

Mixer	Initial investment ($)	Operating costs ($/annum)			
		Degree of mixing			
		1·0	0·8	0·6	0·5
A	10 000	12 000	6000	3000	2000
B	15 000	8 000	4000	2500	1500
C	25 000	5 000	3000	2000	1000

Table 11.2. Data on heating operation*

Degree of mixing	Operating costs ($/annum)			
	Temperature (°F)			
	650	700	750	800
1·0	500	1 000	6 000	10 000
0·8	1 000	1 500	8 000	12 000
0·6	1 500	2 500	10 000	16 000
0·5	2 000	3 000	12 000	20 000

* Initial investment ($). Standard heater. $5000 for temperatures at or below 700 °F. Special heater. $20 000 for temperatures exceeding 700 °F.

Table 11.3. Per cent conversion in primary reactor

Temperature (°F)	650		700		750		800	
Catalyst	1	2	1	2	1	2	1	2
Reactor I_A (%)	30	25	40	30	50	45	60	50
I_B (%)	25	20	30	25	45	40	50	45
I_C (%)	20	15	25	20	40	30	45	40

In general, no problem arises with regard to the reversal of information flow. Thus design can proceed in the direction of mass flow to build up the various design boxes (it is usually more convenient to calculate from reaction forwards). Having produced these boxes as in the example, over a range of 'states', information flow can be reversed. The theory of optimality then

Table 11.4. Primary reactor costs

	Initial cost ($)	Operating costs ($/annum)
Reactor		
I_A	40 000	4 000
I_B	20 000	2 000
I_C	5 000	1 000
Catalyst		
1	—	10 000
2	—	4 000

Table 11.5. Conversion after the clean-up reactor

Conversion in primary reactor	(%)	15	20	25	30	40	45	50	60
Clean-up reactor									
II_A	(%)	30	40	50	60	80	85	90	95
II_B	(%)	45	60	75	85	90	95	95	95
none	(%)	15	20	25	30	40	45	50	60

Table 11.6. Cost of clean-up reactors

Clean-up reactor	Investment ($)	Operating costs ($/annum)
II_A	60 000	10 000
II_B	80 000	20 000
none	0	0

requires that we work in the opposite direction. This is precisely the way we did the initial design. Optimal decisions may be made and the design optimized.

(Tables 11.1 to 11.8 from *Strategy of Process Engineering* by D. F. Rudd and C. C. Watson. Copyright © 1968, by John Wiley & Sons, Inc. Reprinted by permission.)

Table 11.7. Separation costs

Per cent conversion	One large separator		Two small separators	
	Initial cost ($)	Operating cost ($/annum)	Initial cost ($)	Operating cost ($/annum)
30	12 000	2 500	15 000	3 000
40	12 000	3 000	15 000	3 000
45	12 000	4 000	15 000	3 000
50	15 000	4 000	18 000	3 000
60	15 000	5 000	18 000	4 000
75	20 000	6 000	24 000	4 000
80	20 000	6 500	24 000	4 000
85	20 000	7 000	24 000	4 000
90	20 000	7 500	24 000	5 000
95	20 000	8 000	24 000	5 000

Market conditions. The market is rather flexible and the selling price of the pure product is expected to vary with the amount produced by this system. To satisfy internal demands the system must produce at least 15 000 lb/annum. Table 11.8 summarizes the market conditions. A 100% conversion corresponds to 50 000 lb/annum.

Table 11.8. Expected selling price

Production (lb/annum)	Selling price ($/lb)
47 500	3·2
45 000	3·3
42 500	3·4
40 000	3·6
37 500	3·8
30 000	4·6
25 000	5·0
22 500	5·2
20 000	5·3
15 000	5·5

Raw materials. The raw materials used cost 50 000 $/annum.

Table 11.9. Design the blender

Select the optimum design over a range of 'degree of mixing'

Degree of mixing	Mixer type	Operating cost ($/annum)	Optimum 1 ($/annum) Costs
1·0	A	25 600	
	B	24 400	
	C	24 000*	24 000
0·8	A	23 200	
	B	22 800*	22 800
	C	23 200	
0·5	A	21 600*	21 600
	B	21 800	
	C	22 400	

* indicate optimum. The optimum values are carried forward as optimum 1.

Table 11.10. Design the heater

	Reactor inlet temperature (°F)	Degree - of mixing	Optimum cost of heater $0·4(C+0·2I)$	Optimum 1 ($/annum)	Optimum 2 ($/annum) Costs
Standard heater	650	1	600	24 000	
		0·8	800	22 800	22 800
		0·5	1 200	21 600	
Special heater	750	1	4 000	24 000	
		0·8	4 800	22 800	27 600
		0·5	6 400	21 600	
Special heater	800	1	5 600	24 000	
		0·8	6 400	22 800	29 200
		0·5	9 600	21 600	

To identify optimum 2 from the summation of operating costs of the heater and optimum 1, only the lowest costs of this summation are given.

Table 11.11. Design the primary reactor

% conversion after reactor	Reactor type	Operating temperature	Reactor operating costs	Optimum 2 ($/annum)	Optimum 3 ($/annum) Costs
30	A–1	650	8 800	22 800	
	A–2	700	6 400	25 200e	
	C–2	750	2 400	27 600	30 000
40	A–1	700	8 800	25 200e	
	C–2	750	4 800	27 600	
	C–2	800	2 400	29 200	31 600
	B–2	750	4 000	27 600	31 600
50	A–1	750	8 800	27 600	
	B–1	800	6 400	29 200	35 600
60	A–1	800	8 800	29 200	38 000

e is estimate.

Table 11.12. Design the clean-up reactor

% conversion after clean-up reactor	Reactor type	% conversion of input feed	Reactor operating costs	Optimum 3 ($/annum)	Optimum 4 ($/annum) Costs
95	IIA	60	8 800	38 000	46 800
	IIB	60	14 400	38 000	
	IIB	50	14 400	35 600	
	IIB	45	14 400	33 600e	
90	IIA	50	8 800	35 600	44 400
	IIB	40	14 400	31 600	
60	IIA	30	8 800	30 000	
	IIB	20	14 400	28 400e	
	none	60	0	38 000	38 000
30	IIA	15	8 800	28 000e	
	none	30	0	30 000	30 000

e is estimate.

Table 11.13. Design the separators

% conversion after clean-up	Number of separators	Sales revenue 0·4S	Separator costs 0·4(C+0·2I)	Optimum 4 ($/annum)	Optimum 5 ($/annum) Profits
95	1	60 800	4 800	46 800	
	2	60 800	3 920	46 800	10 080
90	1	59 400	4 600	44 400	
	2	59 400	3 920	44 400	11 080
60	1	55 200	3 200	38 000	
	2	55 200	3 040	38 000	14 160
30	1	33 000	1 960	30 000	1 040
	2	33 000	2 400	30 000	

Example 11.3. Optimization of the toluene hydrocracker. This example is included to show the difficulties in resolving real systems and how the cost of a material supplied to the system can alter the selected design.

Objective. Using the process flow diagram, Figure 6.13, select the optimal design for an output of 80 000 t/annum (10 t/h) of 99·96% benzene using reformer hydrogen, 30 wt.% methane. The cost function is of the type

$$U = C_V + dI + mI + \ldots\ i_m I/(1-t)\ \ldots \qquad \text{Table 4.2}$$

which is simplified to $U = C_V + 0·4I$.

Process analysis. An information flow diagram was derived and modified to break the recycle, Figure 11.4. As W_4 (reactor products) is a function of T_3 (vaporizer temperature) the streams T_3 and W_1 were preferred as assumed streams. The system does not decompose easily for dynamic programming and a different approach is attempted.

As the system is required for a specific output it is possible to derive flow rates in terms of some of the variables. As outlined in Example 6.3 the fraction of the feed recycled x_R is given by

$$x_R = (100 - 100C)/C$$

where C is conversion per pass. Neglecting losses the toluene in the feed equals benzene product, i.e. $10 \times 92/78 = 11·8$ t/h. Thus toluene in recycle $= 11·8 x_R = 11·8(100 - 100C)/C$. Allowing for 13% of heavy polymers and recycle benzene, R_b. Total weight of recycle $= 11·8x/(0·87 - 0·01R_b)$.

Gas balances are identified as in the example, p. 60.

Hydrogen balance: $0·7A = 0·128*2 + x_{F.G.}B$
Methane balance: $0·3A = -0·128*16 + (1 - x_{F.G.})B$

where A and B are hydrogen and fuel gas flow rates.

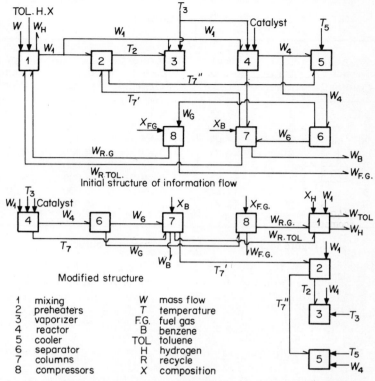

Figure 11.4 Breaking recycle information flow, toluene hydrocracker

$$B = 15 \cdot 1/(7 - 10x_{F.G.})$$
$$A = 0 \cdot 366 + 21 \cdot 57x_{F.G.}/(7 - 10x_{F.G.})$$

The total hydrogen recycled is based on maintaining a 5:1 hydrogen: aromatics molar ratio and is approximated by

$$\frac{10}{x_{F.G.}}\left(0 \cdot 128 + 0 \cdot 128x_R + \frac{0 \cdot 102x_R}{0 \cdot 87 - 0 \cdot 01R_b} - 0 \cdot 7A\right)$$

or

$$\frac{1}{x_{F.G.}}\left(1 \cdot 024 + 1 \cdot 28x_R + \frac{0 \cdot 102x_R}{0 \cdot 87 - 0 \cdot 01R_b} - \frac{15 \cdot 1x_{F.G.}}{7 - 10x_{F.G.}}\right)$$

Thus all the flows are given by expressions such as

$$W_i = f_i(x_{F.G.}; C; R_b)$$

and the information flow may then be reduced to Figure 11.5. T_s; the temperature in separator, and the cost of fuel gas were defined and T taken as T_3 to allow for control of reaction conditions. The variables to be optimized include the following.

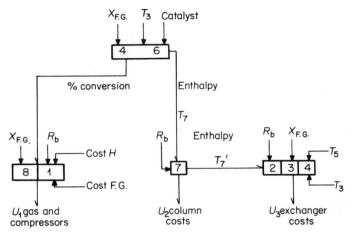

Figure 11.5 Decomposition for specified output

(a) Hydrogen content of fuel gas, $x_{F.G.}$.
(b) Different reactors and catalysts.
(c) Intermediate temperatures on exchanger trains.
(d) Benzene content of recycle toluene. The flowsheet is assumed invariant, other alternatives being examined separately.
(e) The cost of hydrogen is deliberately varied to show how this system is a subsystem of a larger complex and is adjusted by its environment.

The results of the analysis are indicated in Table 11.14. Benzene recycle had been optimized at 4% benzene in recycle toluene from a study of the columns. It is clear that $U_{compressors}$ and $U_{reaction}$ have a marked effect on the system with the cost of hydrogen affecting the decision. The critical design equations are as follows.

(a) Reactor cost:

$$I_R = I_B(Q/Q_B)^{0.65}$$

where Q is capacity and B base case. If constant residense time is the main criterion of conversion then for a specific temperature and catalyst

$$U_{reaction} = I_B f_2(x_{F.G.}; C)$$

(b) Gas cost:

$$U_{comp.} = f_3(x_{F.G.})$$
$$U_{B-A} = f_4(x_{F.G.}, cost_H, cost_{F.G.})$$

An overall model of this critical region may readily be evolved. The effect of hydrogen cost on selection is shown in Table 11.15 which balances reactor costs and fuel gas hydrogen credit. Optimum values are underlined.

Table 11.14. Cost balances for toluene hydrocracker

Per cent conversion	Catalyst	Reactor temperature (°F)	Wt. % hydrogen	Reactor cost	Separator cost	$U_{reaction}$ (£/annum)	U_{column} (£/annum)	$U_{exchangers}$ (£/annum)	$U_{compressors}$ (50£/t)	(70£/t)	Optimum overall cost
90	A–2	1 300	10	650 000	2 880	26 130	12 665	66 016	−35 848	36 326	300 579 (70 £/t)
			20	290 000	2 700	117 000	12 665	67 016	1 374	140 934	171 121 (50 £/t)
			30	50 000	2 180	20 880	12 665	69 516	98 095	310 695	
85	B–2	1 100	10	585 000	2 880	235 200	12 660	66 024	−45 680	36 820	
			20	262 000	2 580	105 900	12 660	68 524	6 557	144 557	
			30	45 000	2 330	18 940	12 660	74 524	97 007	309 007	
85	A–1	1 300	10	520 000	2 880	209 200	12 660	66 266	−45 680	36 820	
			20	232 000	2 580	93 900	12 660	66 216	6 557	144 557	
			30	40 000	2 330	16 940	12 660	67 016	97 007	309 007	
80	A–1	1 100	10	585 000	3 100	235 400	12 656	66 524	−45 124	36 376	
			20	262 000	2 700	105 800	12 656	65 024	7 429	145 429	
			30	45 000	2 300	18 960	12 656	68 524	97 369	309 369	
70	C–1	1 300	10	585 000	3 120	235 248	12 831	68 516	−43 779	37 721	
			20	262 000	2 730	105 892	12 831	66 516	9 631	147 631	
			30	45 000	2 430	18 972	12 831	66 016	98 095	310 095	
75	B–1	1 300	10	455 000	3 100	183 400	12 740	67 516	−44 577	36 923	
			20	207 000	2 700	83 900	12 740	66 016	8 465	146 465	
			30	35 000	2 300	14 940	12 740	65 216	97 713	309 716	

Table 11.15. Effect of hydrogen cost on fuel gas composition

	Values of cost function			
Cost of hydrogen (£/t)	35	50	70	85
Hydrogen in fuel gas; 10 wt. %	65 850	126 900	208 400	270 400
20 wt. %	−28 100	74 900	212 900	315 900
30 wt. %	−80 560	95 924	307 900	473 900
40 wt. %	−21 300			

11.1.4 RIGOROUS APPROACH TO DYNAMIC PROGRAMMING

For a rigorous approach to dynamic programming see Beveridge and Schechter (1970) and Aris (1964). The latter has defined the various terms.

The state of the material on which process is operating is defined by the state variables denoted by the terms $(p_{N+1} \cdots p_1)$. The decision variables are those variables which can be controlled or chosen in the design and operation of any stage and are denoted by the decision vectors $(q_{N+1} \cdots q_1)$. Each stage of the process transforms the input state into an output state in a way dependent on the operation of the stage. Algebraically at the nth stage

$$p_n = \mathcal{T}_n(p_{n+1}; q_n)$$

meaning that given p_{n+1} and q_n we can calculate p_n. Consider the sequence

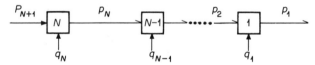

For the final stage if the feed is specified, p_2, then the principle of optimality dictates that the decision vector q_1 is selected so that the objective function y is an optimum for that feed. Denoting the optimum by M_1, then

$$M_1(p_2) = \underset{q_1}{\mathrm{opt}}\{y_1(p_2; q_1)\}$$

and for the last two stages

$$M_2(p_3) = \underset{q_1,q_2}{\mathrm{opt}}\{y_1(p_2; q_1) + y_2(p_3; q_2)\}$$

or as q_1 is selected to optimize y_1 for a given p_2

$$M_2(p_3) = \underset{q_2}{\mathrm{opt}}\{M_1(p_2) + y_2(p_3; q_2)\}$$

or in general

$$M_n(p_{n+1}) = \underset{q_n}{\mathrm{opt}}\{M_{n+1}(p_n) + y_n(p_{n+1}; q_n)\}$$

where $p_n = \mathscr{T}_n(p_{n+1}; q_n)$, $n = 1, 2, ..., N$. Of course this case is extremely simple because no branched streams enter the problem. In the latter case it is not possible to optimize without increasing the number of state variables. Aris (1964) has considered some possible cases which provide an introduction to the greater problem. A simple example is when two branches occur

The maximum returns from the separate branches can be calculated by

$$M_2(p_3) = \max_{q_2} M_1 \{M_1(p_2) + M_1'(p_2') + y_2(p_3; q_2)\}$$

and

$$M_n(p_{n+1}) = \underset{q_n}{\mathrm{opt}}\{M_{n-1}(p_n) + y_n(p_{n+1}; q_n)\} \qquad n = 3, ..., N$$

Branching occurs in practical problems and it is obvious that the solutions used earlier are approximate. But they do represent an attempt to solve the problem where the transform $\mathscr{T}$ is more an informed judgement. Many variables are often negligible in their effect on costs, so frequently such methods are not as inaccurate as precise mathematical solution would indicate. However, the warning must be noted and care taken to ensure the problem has been reduced to straight chain or simple branching.

Example 11.4. To clarify the use of the general formulae a short problem on optimal equipment-replacement strategy has been devised by Beveridge and Schechter (1970).

A process has an item of equipment t years old and it has no salvage value. The costs of operating the item increase with age and the profit of the process is given by

$$p(t) = 25 - t^2$$

If the unit is replaced the cost is 21 and in that year

$$p(0) = 25 - 21 = 4$$

Aspects such as depreciation, etc., are neglected. The decision variable thus has the value 'keep' K or 'replace' R. Find the best set of decisions K or R to maximize the profit over the four years that the process will be operated before its closure. By dynamic programming we may write

$$M_n(t_{n+1}) = \max_{\substack{K \ R \\ n \ n}} \{P_n(t_{n+1}) + M_{n-1}(t_n)\}$$

Starting with one year of operation left, the one-year policy for an item

$$M_1(t_2) = \max_{K_1 R_1} \{P(t_2)\}$$

this value may be calculated assuming different ages of the unit.

Age of unit at previous period t_{N+1}	Profit from n years' operation	N_n, profit for n years' operation
1	$24K$ or $4R$	24
2	$21K$ or $4R$	21
3	$16K$ or $4R$	16
4	$9K$ or $4R$	9

The two-year policy for a three-year old item is given by

$$M_2(t_3) = \max_K\{p(3)+M_1(4)\} \quad \text{or} \quad \max_R\{p(\theta)+M_1(1)\}$$

Since $M_1(4) = 9$ and $M_1(1) = 24$ (both from a keep decision) we have two possible results.

$$K : \sum_{i=1}^{2} P_{(t)} = 16+9 = 25$$

$$R : \sum_{i=1}^{2} P_{(t)} = 4+24 = 28 = M_2(t_3)$$

Evaluating for all cases gives the following.

M_n, profit from n years' operation	Years of process operation left, n			
	1	2	3	4
Age of unit, t_{n+1}, at end of previous period				
1	$24K$	$45K$	$61K$	$73K$
2	$21K$	$37K$	$49(K/R)$	$70K$
3	$16K$	$28R$	$49R$	$65(K/R)$
4	$9K$	$28R$	$49R$	$65R$
5	$4R$	$28R$	$49R$	$65R$

The desired solution is found by entering the list $n = 4$, $t = 2$ and the optimal policy of $KR\ KK$ would yield a profit of 70.

Similar results could be obtained by direct search, and the economy of dynamic programming is more marked as the number of stages, decisions, and initial states increase.

11.2 Linear programming

Consider the linear function

$$y(x_1, x_2, x_3) = ax_1 + bx_2 + cx_3$$

The derivatives are

$$\frac{dy}{dx_1} = a, \frac{dy}{dx_2} = b, \frac{dy}{dx_3} = c$$

There are therefore no values of x_i which make the partial derivatives vanish—a necessary condition for an internal optimum. Since there are no discontinuities in the function, then, provided a, b, c, are constant, the optimum can lie only at a boundary.

Imposing boundary restrictions $0 \leqslant x_i \leqslant 1$

$$i = 1, 3$$

The extreme value will lie at the intersection of the constraining planes. The search is therefore confined to these boundaries and the procedure is termed linear programming. The following example indicates the manual solution of a blending problem.

Example 11.5. We wish to blend three ores A, B, C to form 100 kg of alloy which must contain at least 20% iron, 25% lead, and 48% copper. Compositions and costs of ore are as follows.

Ore	A	B	C
Iron	0·7	0·6	0·0
Lead	0·2	0·1	0·4
Copper	0·1	0·3	0·6
Cost (£/kg)	1000	2000	3000

The objective is the minimum cost blend.

Let x_A, x_B, x_C refer to the weight of ore.

$$x_C = 100 - (x_A + x_B)$$

So eliminating x_C we have

$$\begin{aligned}
\text{cost of alloy, } y &= 1000x_A + 2000x_B + 3000\{100 - (x_A + x_B)\} \\
&= 300\,000 - 2000x_A - 1000x_B
\end{aligned}$$

Constraints are

iron $0·7x_A + 0·6x_B \geqslant 20$

lead $0·2x_A + 0·1x_B + 0·4\{100 - (x_A + x_B)\} \geqslant 25$

Therefore

$$0{\cdot}2x_A+0{\cdot}3x_B \leqslant 15$$
copper $\qquad 0{\cdot}5x_A+0{\cdot}3x_B \geqslant 12$

(i) *Graphical solution.* Plotting the constraints as equalities on a graph, Figure 11.6, illustrates that the permitted region in the (x_A, x_B) plane must lie within these lines. The function $y = 300\,000 - 2000x_A - 1000x_B$ is a family of lines which increase in cost as indicated. The minimum cost is at $x_A = 75$, $x_B = 0$; one of the five vertex.

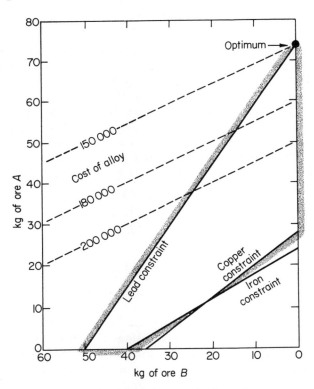

Figure 11.6 Search around a boundary

(ii) *Simplex method solution.* Let us pose the problem as

$$\max 2000x_A + 1000x_B$$

for constraints

$$0{\cdot}2x_A+0{\cdot}3x_B \leqslant 15$$
$$0{\cdot}5x_A+0{\cdot}3x_B \geqslant 12$$
$$0{\cdot}7x_A+0{\cdot}6x_B \geqslant 20$$

So let

$$y = 2000x_A + 1000x_B$$

First modify the constraints to the correct form as follows.

(a) Constraint modification

$$-(0\cdot5x_A + 0\cdot3x_B) \leqslant -12$$
$$-(0\cdot7x_A + 0\cdot6x_B) \leqslant -20$$

(b) Add slack variables x_3, x_4, x_5, to remove the inequalities

$$0\cdot2x_A + 0\cdot3x_B + x_3 = 15$$
$$-0\cdot5x_A - 0\cdot3x_B + x_4 = -12$$
$$-0\cdot7x_A - 0\cdot6x_B + x_5 = -20$$

The initial problem is to obtain a 'feasible solution' which by definition is at a vertex formed by the boundary planes. It we set $n-m$ variables to zero (n variables and m restrictions) and get non-negative values for the remaining variables, this is such a solution. In this case the feasible solution may not be obtained by casual inspection; as when the obvious choices x_A and x_B are zero then x_4 and x_5 are negative which is not permitted. Note on Figure 11.6 how x_A, $x_B = 0$, 0 is not at a boundary.

(c) If the first feasible solution is not at the origin, add further slack variables to make constraints positive when x_A and x_B are zero.

$$0\cdot2x_A + 0\cdot3x_B + x_3 = 15$$
$$0\cdot5x_A + 0\cdot3x_B - x_4 + x_{100} = 12$$
$$0\cdot7x_A + 0\cdot6x_B - x_5 + x_{101} = 20$$

The objective function is now

$$y = 2000x_A + 1000x_B - P(32 - 1\cdot2x_A - 0\cdot9x_B + x_4 + x_5)$$

The penalty function P thus removes the additional slack variables from the solution.

(d) Shorthand algebraic procedures are usually used to avoid writing out lengthy equations. So the above equations are written as an array. The variables not in the basis are set equal to zero and form the first row; variables in the basis then equal the positive values of the first column of numbers, i.e. $x_3 = 15$.

First array:

basis		x_A	x_B	x_4	x_5
x_3	15	0·2	0·3	0	0
→x_{100}	12	0·5	0·3	−1	0
x_{101}	20	0·7	0·6	0	−1
y	0	−2000	−1000	0	0
P	−32	−1·2	−0·9	1	1

The rules of transformation as in the simplex method (Beveridge and

Schechter, 1970) are then applied. The variables to be transformed are decided as follows.

(i) Select the most negative coefficient in the last row of the array if the intention is to maximize (or the most positive coefficient if minimizing), e.g. $-1\cdot2$.

(ii) Based on the ratio of the elements in the first column of number to those in the column selected above, i.e. x_A, choose the row having the smallest positive ratio and remove that row out of the basis. For example, in the first array the ratios are $15:0\cdot2$; $12:0\cdot5$; $20:0\cdot7$, so transform x_A into the basis, and remove x_{100}.

It should be added that for this particular array we wish to remove x_{100} and x_{101} first so we can eliminate P. Once either of these variables are set to zero that variable need not be written in the array.

(iii) Arrow the pivotal column and row and transform as indicated in Table 11.15.

Table 11.15. Rules of transformation in simplex method (Beveridge and Schechter, 1970)

---

(1) Take the left-hand column of variables and top row of variables, and write these in corresponding positions in the new array, with two of the positions interchanged, namely the variable to be added to the basis (from the top row) and the variable to be removed from the basis (from the column), using the following steps.

(2) For convenience draw lines through the row corresponding to the variable being removed from the basis, and through the column corresponding to the variable being added to the basis. Call these the pivotal row and pivotal column, any coefficient being represented by p_r or p_c and the intersection of the lines being the pivotal element, p_p.

(3) Calculate the reciprocal of the pivotal element $(1/p_p)$ and place in the corresponding position in the new array.

(4) For each coefficient in the pivotal row calculate p_r/p_p and place in the appropriate location in the new array.

(5) For each coefficient in the pivotal column calculate $-p_c/p_p$ and relocate.

(6) For any unfilled location calculate

$$p - \frac{p_r p_c}{p_p}$$

and relocate. p_r is the value in the pivotal row in the same column as the element p, p_c is the value in the pivotal column as the element p, and p_c is the value in pivotal column in the same row as p.

Second array:

basis		x_B	↓ x_4	x_5
x_3	10·2	0·18	0·4	0
x_A	24	0·6	−2	0
→x_{101}	3·2	0·18	1·4	−1
y	48 000	200	−4000	0
P	3·2	−0·18	−1·4	1

Transform x_{101} and x_4. Note that when x_{100} and x_{101} both equal zero that P is eliminated, i.e.

$$P*(x_{101}+x_{102}) = P*0 = 0$$

so do not write in this row.

Third array:

		x_B	$\downarrow$ x_5
$\rightarrow x_3$	9·286	0·13	0·286
x_A	28·57	0·857	−1·43
x_4	2·32	0·13	−0·71
y	57 140	714	−2860

All elements in the last row must be positive at the solution, i.e. no value to maximize.

Fourth array:

		x_B	x_3
x_5	32·6	0·45	0·35
x_A	75·0	1·117	0·5
x_4	25·5	0·455	0·25
y	150 000	1900	10 000

This is the solution

$$y = £150\ 000 \text{ at } x_A = 75·0, \qquad x_B = 0·0$$

The values of x_5 and x_4 have no meaning with respect to the blend. They only indicate the mathematical location of the vertex in space.

As an exercise the reader may attempt to minimize

$$y = 300\ 000 - 2000x_A - 1000x_B$$

Proceed as above until x_{100} and x_{101} are eliminated and then minimize (see note (i)). Continue until all elements in the last row are negative. Most users of linear programming concentrate on obtaining the constraints in the correct form and evaluating the problem by computer using an available program, i.e. optisep (Siddall, 1970). The task of the process engineer is mainly in recognizing the problem as one solvable by linear programming.

The technique is particularly useful in operations planning when the system to be resolved is flexible. This explains its original extensive use by the petroleum-refining industry (Symonds, 1956) on crude oils. Thus, if we wished to process crude oil through a still, cracker, etc., and the products were gasoline, naphtha, jet fuel, fuel oil, the optimal solution would be the particular feasible solution which maximizes the value of the objective function

$$-8x_1 - x_1 - 2x_{12} + 18x_3 + 10x_5 + \dots$$
$$\text{crude \quad still \quad cracker \quad gasoline \quad naphtha}$$

where costs are negative and realizations are positive. The integers in this case represent £/kg and the variables kg/d. Furthermore there could be constraints which limit the production, say

$$\text{gasoline } x_3 \leqslant 20\ 000$$
$$\text{naphtha } x_5 \leqslant 10\ 000$$
$$\text{fuel oil } x_{18} \leqslant 6\ 000, \qquad \text{etc.}$$

These constraints could be availability of raw materials or product requirements.

Formulating constraints is an important problem of the linear programmer. Sometimes this may involve adaption of the main program. Allen (1971) gives a useful example to illustrate the problem of plant shut-down.

Example 11.6. Suppose an impure product is made to 70% yield in a reactor, followed by purification in a distillation column the maximum feed rate through which is 120 000 kg/d. The yield of product from distillation is 92% and the maximum requirement of pure product is 2 700 000 kg over a 30-day period. The distillation unit will be shut down for maintenance for 10 days in the period and during this time 300 000 kg of impure product will be put to storage.

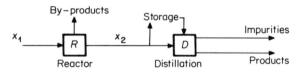

(a) Production of pure product $= 0{\cdot}92x_2$ kg/d
Maximum requirement $= 90\ 000$ kg/d
Constraint to ensure that production does not exceed requirements

$$0{\cdot}92x_2 \leqslant 90\ 000$$

(b) Reactor product $= 0{\cdot}7x_1$
Thus balance x_2 and material to storage 10 000 kg/d
Balance constraint is

$$-0{\cdot}7x_1 + x_2 = -10\ 000$$

Example 11.7. Optimum distribution of products. Let us try a practical example of planning distribution of a companies' products to market. This will be restricted to three interrelated products for ease of presentation. Let us assume that these are produced in two works, distributed internally as required for production, and sold to five markets as shown in Figure 11.7. Information about the capacity of the works and markets during the period under review and transport costs are indicated in Table 11.16. As the aim is

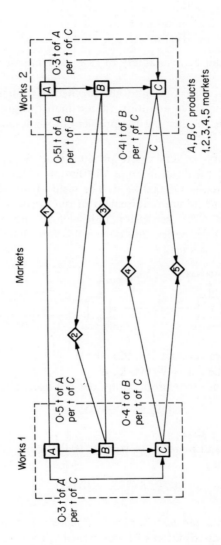

Figure 11.7 Distribution of company products

Table 11.16. Cost and capacity information

Products	Identi-fication number	Works capacity of plant	Variable costs	Transport costs	Market identification number	Capacity	Selling price
A	1	250	10	2	1 →	(1) 100	15
	2	300	9	1	1 →		
B	1	300	5	2	2	(2) 200	10
				1	3		
	2	300	6	1	2	(3) 300	11
				2	3		
C	1	250	7·5	1	4	(4) 400	12
				1	5		
	2	350	7	2	4	(5) 500	13
				2	5		

to maximize the companies' gross margin we must include information on selling price and variable costs.

The first task is to number the arrays and $C_{1,4}$ relates to product C at works 1 being sold in market 4. Mathematical relationships must now be written defining the operation of this company.

Capacity (6 relationships): as indicated earlier a capacity restriction may be written as

$$A_1 \leqslant 250$$

(for other units see linear programming matrix, Table 11.17).

Sales (5 relationships):

$$A_{1,1} + A_{2,1} \leqslant 100$$

means that the total quantity sold cannot exceed the requirements of the market.

Product balance (6 relationships): movement of any product within the works must be in balance, with availability and outside sales

$$B_1 = 0{\cdot}4C_1 + B_{1,2} + B_{1,3}$$

or expressing constraint in correct form

$$0{\cdot}4C_1 + B_{1,2} + B_{1,3} - B_1 = 0$$

Gross margin: this is based on the contributions as follows.
Sales contributions:

$$\{(15-2)A_{1,1} + (15-1)A_{2,1} + (10-2)B_{1,2} + (10-1)B_{1,3}$$
$$+ (11-1)B_{2,2} + (11-2)B_{2,3} + (12-1)C_{1,4} + (12-1)C_{1,5}$$
$$+ (13-2)C_{2,4} + (13-2)C_{2,5}\}$$

Variable costs:

$$(10A_1 + 9A_2 + 5B_1 + 6B_2 + 7 \cdot 5C_1 + 7C_2)$$

and the gross margin to be optimized is given by (sales contributions—variable costs).

All the constraints and gross margin are listed in Table 11.17, the linear programming matrix for this problem. The solution to the matrix is readily found from a linear programming computer routine. We thus have a target level within the period. For instance, the model can be used when production levels are not met or in forward planning. It is emphasized that it must be kept up to date to be of value.

The above example was chosen for simplicity, and similar techniques can be used to evaluate the optimum production pattern for a works, either existing or planned. For real systems the solution of problems becomes hampered by size. Lasdon (1970) indicates that problems of up to 4095 rows and many more variables have been solved using the revised simplex method. For larger problems he shows how an exploitation of system structure by direct and decomposition techniques is required.

11.3 Optimization of real systems

The mathematical model of a system can be written as

$$\max_q \{\phi(p;q) \mid g(p;q) \geqslant 0\} = \phi^*(p) = \phi(p, q^*)$$

where ϕ is the objective function whose value is determined by the decision variable vector q, and the parameter vector p. The various limitations are indicated by the vector of constraint limitations g. The decision variables are adjusted using an optimization algorithm to obtain the best value of the objective function ϕ^* for a parameter vector p which describes the particular case being optimized. Corresponding to ϕ^* there are one or more points, q^*, where $\phi = \phi^*$.

Hughes (1969) has described the operation of CHEOPS which contains a number of algorithms which solve any problem formulated in this fashion. Table 11.18 shows the usable problem size for the optimization algorithms used. As will be noted, the objective function is frequently non-linear. The references in the table provide an introduction to this problem and Box *et al.* (1969) describe numerous non-linear optimization techniques.

The problem in use of such systems is not only one of computation cost but also of formulation time and data organizing. It is necessary to keep the model as simple as is consistent with the problem. Thus, if a heat exhanger is not a key item (it rarely is), use overall heat transfer correlations at the optimization stage. Remember cost information is rarely accurate. Do not specify all components. As in distillation, some components are the 'key'

Table 11.17. Linear programming matrix

Variables	A_1	B_1	C_1	A_2	B_2	C_2	$A_{1,1}$	$A_{2,1}$	$B_{1,2}$	$B_{1,3}$	$B_{2,2}$	$B_{2,3}$	$C_{1,4}$	$C_{1,5}$	$C_{2,4}$	$C_{2,5}$	
capacities:																	
I, product A	1																⩽ 250
II, product A				1													⩽ 300
I, product B		1															⩽ 300
II, product B					1												⩽ 300
I, product C			1														⩽ 250
II, product C						1											⩽ 350
sales:																	
market 1							1	1									⩽ 100
market 2									1		1						⩽ 200
market 3										1		1					⩽ 300
market 4													1		1		⩽ 400
market 5														1		1	⩽ 500
products:																	
I, product A	−1	0·5	0·3				1										= 0
II, product A				−1	0·51	0·3		1									= 0
I, product B		−1	0·4						1	1							= 0
II, product B					−1	0·41					1	1					= 0
I, product C			−1										1	1			= 0
II, product C						−1									1	1	= 0
Gross margin	−10	−5	−7·5	−9	−6	−7	13	14	8	9	10	9	11	11	11	11	maximize

to the problem. Try to use a good base case and study marginal changes, preferably isolating these key areas. Unit subroutines which involve many iterations for their solution are undesirable within the mass balancing routines and simultaneous solution is desirable. In some cases one should not automatically go to a packaged process simulator to evaluate alternatives but spend time in formulating the mass balance routine as a one-off job. Remember that implicit constraints may arise which are limitations on values of items in the process flow but are not optimization parameters, i.e. liquid below its boiling point.

Table 11.18. Problem size for optimization algorithms

	L.P.	M.A.P.	G.P.	P.P.N.L.	D.A.
Objective function	L.	N.L.	N.L.	N.L.	N.L.
Decision variables					
linear	101	500	100	600	40
non-linear		100	—	60	
Bounds			200	120	80
Constraints	}2000	}1000	}200	}2000	20
Equations					—

L. is linear; N.L. is non-linear; L.P., linear programming; M.A.P., Griffith and Stewart (1961); G.P., gradient projection (Rosen, 1960); P.P.N.L., Ornea and Eldredge (1965); D.A., direct ascent (Singer, 1962).

The problem is usually studied as deterministic, but this is itself an approximation. This should be remembered when tackling the problem and be reflected in the desired accuracy of solution. Key parameters should be varied to optimistic and pessimistic values before finalizing on the best solution.

12 Simulation

12.1 Overall design

The complete simulation of an overall design is an important aim. Whether this aim is economically sound is difficult to answer and requires further analysis of the activities of the project network (Mackenzie, 1967). To reduce the extent of the analysis the information flow relating to process charts, specifications, equipment draughting, layout, and pipework is indicated in Figure 12.1. This is necessarily incomplete, as otherwise it would require indicating every item of equipment, such is the complexity. It shows how the development of these aspects is a cyclic exercise taking several weeks. Examination of recycle of information shows which sections require extensive recycle and which, apart from an initial loop, only require recycle for updating purposes. Figure 12.2 illustrates the main flows for the main design activities classified as the following.

(1) Process flow.
(2) Process specification.
(3) Mechanical specification.
(4) Detailed draughting of equipment.
(5) Layout.
(6) Pipe routing.
(7) Pipe detailing.
(8) Fabrication of pipework.
(9) Fabrication of equipment.

This omits in particular civil, structures, and instrumentation.

If this is analysed, it may be seen that the sections may be integrated together into larger unit blocks as in Figure 12.3. This diagram is hardly a surprise as it duplicates many aspects of project organization, thus showing how these are based on reducing information flow between sections. The sections may be considered separately for economic justification.

Certain of the tasks are now carried out by computer. Probably the most successful are the integrated pipework design systems, such as ISOPEDAC (Daniels, 1971). Such systems provide information for estimating, detailing,

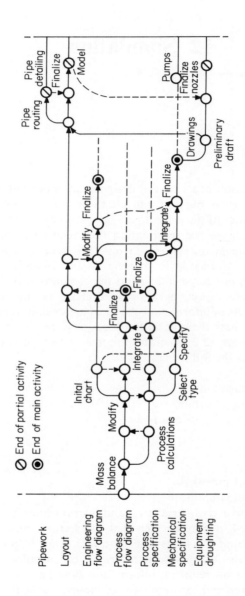

Figure 12.1 Network of project—selected information flow

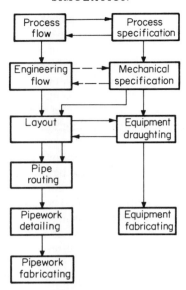

Figure 12.2 Simplified project structure

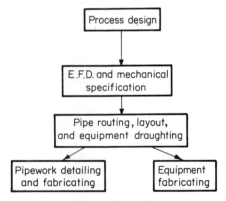

Figure 12.3 Design packages for the computer

costing, and control of pipework. Cost savings of up to 60% and increases in productivity between 5 and 10 to 1 have been reported at the detailing and material listing stage. Additional savings result from the automatic production of bill of costs, fabrication schedules, and materials control.

Whether implementing other activities can be as successful is difficult to forecast. The process design stage is discussed in detail in a following section. Mechanical specification is detailed in Section 8.3. Savings arise from the ability to carry out complex calculations quickly and to call up advanced routines for design of items of equipment. Costs arise in developing programs and machine time. The computer is involved, as otherwise improved sizing techniques could not be used, but this is generally in a partial rather than an overall role. Thus design of individual items of equipment is readily handled by computer. Process specification is commonplace and, as codes of practice are increasingly translated from graphical to algebraic form, so mechanical specification can be handled at the same time. Programs have been devised which completely specify the equipment and produce this information in the form of a specification data sheet. But nozzle connections have to be added separately off engineering flow diagrams and the sketch is normally added by hand.

Full mechanical specification, including drawings, presents greater problems and, although sketches can now be provided using digital plotters, the complete design requires the use of computer graphics. It would not appear to be economic at the present time because of the minor savings.

Similarly with regard to the engineering flow diagrams or engineering line diagrams one has strong reservations as to the viability of a computer technique. To generate the diagrams requires considerable interaction with a computer graphics section and there is a tendency to use the computer as a draughting board. Unfortunately the draughting effort on such drawings is only small, so the saving in cash is small. On the other hand, there is no doubt that, if a design package for mechanical specification is to be constructed, it must be integrated with engineering flow diagrams and it would be a more cumbersome procedure were data input by typewriter.

A further gap occurs at the layout stage. The number of alternative actions in simply locating 8 blocks relative to one another is 8! or 40 532. Thus on a computer it is essential to use graphics in a interactive environment to synthesize the plot plan. The savings arise from optimizing piping costs as the layout progresses (Bush and Wells, 1972). However, it would seem that a modelling stage will arise intermediate between the layout and pipe detailing stage.

Referring to Figure 12.2, if it is accepted that there must be a break in the overall computer design concept at the layout stage, for modelling purposes, there is less value in equipment drawing by computer because the two are closely integrated.

Another aspect should be noted. At present the detailed design is often carried out in the offices of the manufacturer or licensor. Thus there is a break in the overall chain. A comprehensive simulator does not readily permit any change to take place in the vessel as designed, because such change could interfere with activities already continued by the simulator in the next stage of the design. It could well be more expensive to change such plans than any saving by buying a standard item. This is unfortunate as a natural gap occurs when bid-tabulations are considered. Similarly detailed drawing is still necessary by the manufacturer before fabricating equipment and a further break is introduced.

To conclude this analysis it would seem desirable and eventually economic to attempt to produce a package for process design, a package for mechanical specification and engineering flow diagram (mechanical design), a costing package, and facilities for linking the packages. The advantage of the latter is that it is essential to optimize the process design at an early stage and this can conveniently be carried out by inputting cost information from the mechanical specification stage. This would permit the licensor of a process to produce an optimized process package for input to the mechanical design package. The output from the mechanical design package would be completed flow diagrams, data sheets, and preliminary costs. This would feed a manual layout and bid assessment stage. Detailed draughting would be carried out by manual means, but pipework design following initial routing would probably be computerized.

The significant aspects for the process engineer must be an increased responsibility for mechanical design and the complete specification of equipment. Eventually the overall simulator will be achieved, but the initial aim should be to perfect the initial process and mechanical design packages and the pipework packages. Of course, where the computer can result in savings outside such packages it should be used.

For further information on computer hardware see Daniels (1971) or Taylor (1971).

12.2 Process design simulation

The general aim of a computerized process design system is that it should carry out mass and energy balances, sizing of equipment, costing, and optimization. In extent this will be related to the process engineering stage which the project has reached.

At the preliminary stage, when approximate details will suffice, it is possible to avoid the necessity for complex models; balancing of recycles and the calculation of the essential operating parameters may be dealt with in one operation. Andrew et al. (1971) have reported an I.C.I. package which effectively solves the problems. The system works as follows. First a block flow diagram is drawn. A model is selected for each unit in the diagram. It has been found that the simple mass balance subroutines provided with the system are adequate for modelling most of the operations with the exception of any reaction sections for which special models have often been required. Streams are named and interconnections defined. Recycle streams must be recognized and balanced to an appropriate accuracy. An iterative algorithm is used for balancing the recycles. The main blocks are not specified other than as a cost of the operation. Thus DISTC costs a distillation system, and it is the system only which is sized, using information available from similar systems previously established. This greatly simplifies .the subroutines. The capital cost generated is for the installed unit and optimization is carried out by a series of sensitivity studies with respect to changes in the process parameters. As reported, the program at present is more concerned with capital cost estimation and, to fulfil the general requirement, would have to be expanded to incorporate profitability.

The program is not capable of synthesis of the flowsheet and in fact only limited advances have been made in this field. Siirola *et al.* (1971) have developed a method which assigns to the computer a summary of the performance abilities of existing equipment. The computer is programmed to select its own consistent equipment arrangements using heuristic selection procedures. This technique is valuable in producing a variety of feasible processes, some of which had not been identified previously, developing design experience, and could lead to rules of process invention which would prove useful to the process engineer.

At the budget stage a more complex system is required. A problem is how to deal with the network of interconnecting blocks when recycle loops exist. The simple techniques for breaking the loop by visual examination or, say, price rises are not so practicable when evaluating a complex system using a sophisticated generalized program. Thus programs for ordering are required. As each block with its own internal recycles may take a long time to evaluate, considerable effort has been devoted to devising efficient techniques for this purpose.

Several major simulation programs are in general use, including PACER, GEMCS, CHESS, SPEED-UP, CPCSYS, CHIPS, NETWORK 67, CHEOPS, FLOWPACK, and an early review is given by Evans *et al.* (1968) and Henton and Johns (1971).

These programs are concerned with steady state simulation, but increasingly systems are being developed for dynamic simulation; DYNSYS, CSL. Essentially all these systems consist of an executive program which processes information between modules including data storage and supply. A module can be a unit or it might simply add input flows together and compute desired outputs, i.e. a junction of pipes. Each module has its own format for supply of information, aspects such as location of flowsheet number and type of unit being standardized. Similarly as in normal mass balance calculations the flow of each component is listed in the same location for each stream.

Obviously the engineer will use the facility available to his company and detailing a particular system would serve little purpose in a guide of this nature. Some systems stop at mass balances and set operating parameters and are more suitable for the simulation of existing plant. Others size units and can evaluate costs. Thus modules for distillation can range from a simple mass balance routine involving no equilibrium data to a major program which evaluates the column and ancillaries. Costs are normally generated from indices. If the latter also incorporate cost information stored from previous exercises within the company, then such an evaluation should be accurate enough for budgeting purposes.

At the control stage it is possible to use the above simulators for mass and energy balances, but the opportunity exists for re-sizing individual equipment using procedures available at licensors/manufacturers. Thus the facility might size, say, a heat exchanger for budgeting purposes by a simulator using a general sizing procedure, but then obtain a more accurate design with improved information from a specialized manufacturer once the tender is obtained. The cost information for this stage is obtained from bid-tabulation rather than the simulator, but a check should be made that economic design errors have not been perpetrated on the simulator should the costs differ greatly. One aspect which may require further effort by the process engineer is to check whether optimization of individual items has been carried out or whether general formulae and 'rules of thumb' which do not take into account financial aspects such as grants and allowances have been used by the simulator. Had improper suboptimization of an interior component taken place, such an error is relatively easy to make on a simulator. For further reading see the books by Crowe *et al.* (1969) and Himmelbau and Bischoff (1968).

Example 12.1. Process simulation of the toluene hydrocracker. As the process flow of this system has been discussed earlier, Example 12.1, it may assist the reader to indicate briefly how the GEMCS system (Johnson, 1971)

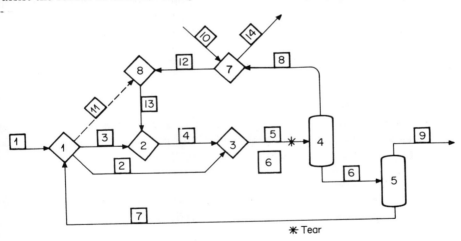

Figure 12.4 Information flow diagram, toluene hydrocracker

can be used to evaluate mass balances. Figure 12.4 shows an information flow diagram of the system. The modules used are indicated in Table 12.1. The system converges easily and no modification as in Figure 11.4. was required as heat balances were not calculated.

Only the short modules 2 and 8 were written specially for the system.

Table 12.1. Modules for toluene hydrocracker

Number	Type	Description
1	JUNCO1	sums toluene feed 1 and recycle toluene 8; outputs aromatic flows
2	REACO3	converts all toluene input to benzene and methane
3	JUNCO1	adds in unconverted benzene 2
4	SEPAO1	removes gas
5	SEPAO1	removes benzene
6	CONTL1	measures convergence of stream 7
7	JUNCO5	adds hydrogen feed and separates fuel gas
8	SETS H2	sets hydrogen:aromatics ratio

The calculation order was readily obtained by inspection. The computer used was an I.C.L. 1900 series, 14K storage required and mill time 70 s. The case study is available from the author.

In conclusion one must add that the user of process simulators can often end up taking longer to evaluate mass balances and size equipment than by conventional means including free standing programs. Further use after the plant is installed often does not result in any cost savings. However, as the systems continue to improve, familiarity with this approach increases in importance. It is emphasized that only by use of the systems can an appreciation of the advantages be made.

12.3 Economic simulation

12.3.1 DESIGN MODEL

By adding costing and reporting modes to the process model economic simulation is possible. Listing should be in three areas.

(a) As a detailed breakdown of the capital cost.

(b) In detail for optimization.

(c) As a profit and loss account of the plant.

The detailed breakdown of the capital cost is useful for control purposes during construction but, after this, only applies for post mortem and future design.

The optimization format is modified by information during production and fresh sensitivity studies can be carried out. The model is useful during commissioning and when increasing capacity. The profit and loss account is generated by less information than required for design and the model is modified accordingly. Note that allowance must be made for heat and material losses at this stage and the model must produce correct costs according to variation in throughput. Variable selling price is covered by a separate selling mode. The initial model sets preliminary plant standards which are made firm as improved information is available from actual costs. Variance in performance is obtained by a comparison of actual and standard costs.

12.3.2 CORPORATE MODEL

The plant model may then be integrated in the corporate model of a larger cost centre. This means further change in the model, as business planning is orientated towards what can be sold. The information flow is from sales to production in the form of products which set material flow. Ideally the larger cost centre is the works and the model is used for forward planning studies; on the phasing in or out of plants, expansion of utilities, sensitivity studies, etc. It is possible to consider the effect of varying parameters on a larger system and transfer prices exert less influence. Complete honesty over consumption is vital and figures should be measured. This must include, say, office heating, air consumption, water, and other minor costs. Thus using the model the effect of closure of a unit should show at all the cost centres concerned. The overall effect is then achieved by summing the centres. Similarly the effect of buying in different materials, by-product sales, expansion, etc., can be measured.

Eventually the model increases in scope to cover the companies' operations and by this stage becomes a linear programming exercise, ideally upgraded further with aspects such as working capital, shares, etc.; but this is getting somewhat remote from process engineering.

No details of such models are given in this section. The essential aspect is that all financial balances are capable of algebraic solution and production of such a model requires application of the same techniques as used for mass balances. The only feature which adds complexity is to ensure due allowance for time value of money.

References

Abrams, H. J., 1970, *Chem. Engr.*, September, 252.

Allen, D., 1971, *Brit. Chem. Eng.*, **16**, August, 720.

Amundson, N. R., Aris, R., Kalman, R. E., and Lapidus, L., 1960, 'Applied maths in chemical engineering', *Am. Inst. Chem. Eng.*, *Spec. Lect. Ser.*, Washington D.C.

Andrew, S. M., Swann, W. H., and Wray, G. P., 1971, *On-line Computer Methods Relevant to Chemical Engineering*, British Computer Society, p. 21.

Aris, R. F., 1964, *Discrete Dynamic Programming*, Blaisdell, New York.

Armistead, G., 1959, *Safety in Petroleum Refining*, Simmonds.

Bauer, R. L., and Orr, C. P., 1954, *Chem. Eng. Progr.*, **50**, June, 312.

Bauman, H. C., 1964, *Fundamentals of Cost Engineering*, Reinhold.

Bellman, R., 1957, *Dynamic Programming*, Princetown University, Princeton, New Jersey.

Betts, G. G., 1972, *Chem. Engr.*, February, 78.

Beveridge, G. S. G., and Schechter, R. S., 1970, *Optimisation, Theory and Practise*, McGraw-Hill, New York.

Box, M. J., Davies, D., and Swann, W. H., 1969, *Non-linear Optimization Techniques*, Oliver and Boyd, Edinburgh.

British Standards Institution, 1953, *B.S. 974: 1953*.

British Standards Institution, 1964, *B.S. 1646: 1964*.

Bush, M. J., and Wells, G. L., 1971, *Brit. Chem. Eng.*, **16**, April, 325.

—— 1972, *Design, Decision and the Computer*, Inst. Chem. Engrs. Symp. Ser., No. 35, 2:15.

Capital Cost Estimation, 1969, Institution of Chemical Engineers, London.

Cavett, R. H., 1963, *Petrol. Inst.*, *Preprint 04–63*.

Cheers, R. F. C., and Furman, T. T., 1971, *Brit. Chem. Eng.*, **16**, 478.

Christensen, J. H., 1970, *Am. Inst. Chem. Engrs.*, 16, No. 2, 177.

Computers in Engineering Design, Vol. II, 1966, University of Michigan, Ann Arbor, Michigan.

Crowe, C. M., Hamielec, A. E., Hoffman, T. W., Johnson, A. I., Shannon, P. T., and Woods, D. R., 1971, *Chemical Plant Simulation*, Prentice-Hall, New Jersey.

Cutts, G. C., and Wells, G. L., 1971, *Chem. Proc. Eng.*, **52**, April.

Dagnall, B. D., 1965, *Brit. Chem. Eng.*, **10**, 760.

Daniels, P. T., 1971, *On-line Computer Methods Relevant to Chemical Engineering*, British Computer Society, p. 46.

—— 1971, *Chem. Engr.*, August, 297.

Elliott, D. M., and Owen, J. M., 1968, *Chem. Eng.*, November, 377.

Evans, L. B., Steward, D. G., and Sprague, C. R., 1968, *Chem. Eng. Progr.*, **64**, No. 4, 39.

Fawcetts, H. H., and Wood, W. S., 1965, *Safety and Accident Protection*, Wiley, New York.

French, K. H. V., 1959, *Crystal Processes for Pure Benzene*, Benzole Producers, London.

Furman, T. T., and Cheers, R. E. C., 1971, *Brit. Chem. Eng.*, **16**, 478.

Generaux, R. P., 1937, *Ind. Eng. Chem.*, **29**, 385.

Gerster, J. A., 1963, *Chem. Eng. Progr.*, **59**, No. 3, 35.

Gregory, S. A., 1968, *Chem. Engr.*, April, 79.

Griffith, R. E., and Stewart, R. A., 1961, *Magn. Sci.*, **7**, 379.

Guerveri, G., 1969, *Hydrocarbon Processing*, **48**, 137.

Hackney, J. W., 1965, *Control and Management of Capital Projects*, Wiley, New York.

Hand, W. E., 1958, *Chem. Eng. Progr.*, **37**, 331.

Happel, J., 1958, *Chemical Process Economics*, Wiley, New York.

Harary, F., 1959, *J. Math. Phys.*, **38**, 104.

Henlay, E. J., and Staffin, H. K., 1963, *Stagewise Process Design*, Wiley, New York.

Henry, J. D., *et al.*, 1969, *Purification of Materials* (Ed. M. Zieff), Dekker, New York.

Henton, J. E., and Johns, W. R., 1971, *DISCOP Conf.*, Gyor, September.

Himelbau, D. M., and Bischoff, K. B., 1968, *Process Analysis and Simulation*, Wiley, New York.

Hooke, A., and Jeeves, T. A., 1969, *Computer Machines*, **8**, 312.

Hughes, R. R., 1969, *Chem. Eng. Education*, Summer, 113.

Johnson, A. I., 1970, *G.E.M.C.S. Manual*, McMaster University, West Ontario.

Lasdon, L. S., 1970, *Optimization Theory for Large Systems*, Macmillan, London.

Lee, W., and Rudd, D. F., 1966, *Am. Inst. Chem. Engrs.*, **12**, 1184.

Lewis, W. K., and Matheson, G. L., 1932, *Ind. Eng. Chem.*, **24**, 494.

Lyle, O., 1947, *Efficient Use of Steam*, H.M.S.O., London.

McCabe, and Thiele, E. W., 1925, *Ind. Eng. Chem.*, **17**, 605.

McKay, D. L., 1969, in *Purification of Materials* (Ed. M. Zieff), Dekker, New York.

Makenzie, G., 1967, *Chem. Engr.*, June, 625.

Merrett, A. J., and Sykes, A., 1966, *Capital Budgeting and Finance*, Longmans, London.

Meyer, A. (Ed.), 1956, *Symp. Monte Carlo Methods*, Wiley, New York.

Miller, P. A., 1963, *Chem. Engr.*, September, 226.

Mitten, L. G., and Nemhauser, G. H., 1963, *Chem. Eng. Progr.*, **59**, January, 1952.

Nagiev, M. F., 1964, *The Theory of Recycle Processes in Chemical Engineering*, Macmillan, New York.

Ornea, J. C., and Eldredge, C. G., 1965, *Inst. Chem. Engrs. Symp.*, June, Paper 4.15.

Perry, J. H. 1963, *Chemical Engineers' Handbook*, McGraw-Hill, New York.

Rase, H. F., and Barrow, M. H., 1957, *Project Engineering of Process Plants*, Wiley, New York.

Rippin, D. W. T., 1969, *Chem. Engr.*, July, 284.

Rosen, J. B., 1960, *J. Soc. Appl. Maths.*, **8**, 181.

Ross, R. C., 1971, *Chem. Eng.*, **78**, 153.

Rudd, D. F., and Watson, C. C., 1968, *Strategy of Process Engineering*, Wiley, New York.

Sargent, R. W. M., and Westerberg, A. W., 1964, *Trans. Inst. Chem. Engrs.*, **42**, 190.

Saunders, and Brown, 1934, *Ind. Eng. Chem.*, **26**, 98.

Schweyer, H. E., and May, F. P., 1962, *Ind. Eng. Chem.*, **54**, No. 8, 46.

Sidall, J. N., 1970, *Optisep*, McMaster University, West Ontario.

Siirola, J., Powers, G. J., and Rudd, D. F., 1971, *Am. Inst. Chem. Engrs. J.*,

Simon-Carves, 1970, *Freshwater by Desalination*, Simon-Carves, Stockport.

Singer, E., 1962, *Chem. Eng. Progr. Symp.*, No. 37, 62.

Steward, D. G., 1965, *J. Soc. Ind. Appl. Math.*, **132**, 345.

Symonds, G. H., 1956, *Ind. Eng. Chem.*, **48**, No. 3, 394.

Taylor, F. E., 1971, *On-line Computer Methods Relevant to Chemical Engineering*, British Computer Society, p. 1.

Tayyabkhan, M., and Richardson, T. C., 1965, *Chem. Eng. Progr.*, **61**, January, 78.

Thiele, E. W., and Geddes, R. L., 1933, *Ind. Eng. Chem.*, **25**, 289.

Wells, G. L., 1971, *Inst. Fuel*, **44**, 606.

Westerberg, A. W., and Edie, F. C., 1971, *Chem. Eng. J.*, **2**, No. 2, 114.

Wilde, D. J., 1964, *Optimum Seeking Methods*, Prentice-Hall, New Jersey.

Williams, T. J., 1961, *Systems Engineering*, McGraw-Hill, New York.

Winkle, M. V., and Todd, W. G., 1971, *Chem. Eng.*, **78**, 136.

Working Party on Plant Layout, 1973, *Rept.*, *Inst. Chem. Engrs.*, I.T.C., to be published.

Index